C000066492

Catalogue of Discordant Redshift Associations

HALTON ARP

Apeiron
Montreal

Published by C. Roy Keys Inc.
4405, rue St-Dominique
Montreal, Quebec H2W 2B2 Canada
http://redshift.vif.com

© C. Roy Keys Inc. 2003

First Published 2003

National Library of Canada Cataloguing in Publication

Arp, Halton C., 1927-
    Catalogue of discordant redshift associations / Halton Arp.

Includes bibliographical references and index.
ISBN 0-9683689-9-9

    1. Galaxies--Catalogs. 2. Galaxies--Clusters. 3. Quasars. 4. Red shift. I. Title.

QB857.A757 2003   523.1'1   C2003-902604-3

Cover design by

Cover Image

The cover picture shows a Hubble Space Telescope image of the disturbed, low-redshift galaxy NGC 4319 connected by a thin filament to the high redshift quasar/AGN, Mrk 205. In 1971 a photograph with the Palomar 200-inch telescope discovered this as a somewhat broader connection. Despite vigorous contentions to the contrary, further photographs have confirmed it (e.g., see Fig. 13 in Introduction).

In 2002 an organization associated with NASA issued a press release with an HST picture, claiming disproof of the connection. However, many amateur astronomers processed the same picture and showed the connection clearly. One of these pictures, processed in false color by Bernard Lempel, is shown here on the cover. It is interesting that the smaller aperture Space Telescope does not show the broader connection as well as large-aperture ground-based telescopes, but its higher resolution shows for the first time a narrow filament inside that connection. In addition, the contouring in the processed picture emphasizes the disturbed nature of the ejecting galaxy and the alignment of the inner and outer cores of NGC 4319 and Mrk 205 and the filament, a circumstance that clearly precludes an accidental background projection.

Even though this latest evidence has still not been acknowledged by mainstream astronomy, it is very satisfying to me in view of the fact that the Kitt Peak National Observatory 4-meter image of the connection was shown on the cover of my first book, *Quasars, Redshifts and Controversies*. The explosive, ejecting, X-ray nature of Mrk 205 was shown on the cover of my next book, *Seeing Red: Redshifts, Cosmology and Academic Science*. Now the conclusive, thin aspect of the connection is shown by Space Telescope and furnishes an appropriate introduction to this *Catalogue*, which presents much further evidence for the ejection of active, high-redshift objects from lower-redshift galaxies.

Table of Contents

Introduction

The Fundamental Patterns of Physical Associations

Empirical evidence which is repeatable forms the indispensable basis of science. The following *Catalogue of Discordant Redshift Associations* applies this principle to the problem of extragalactic redshifts. The *Catalogue* entries establish unequivocally that high redshift objects are often at the same distance as, and physically associated with, galaxies of much lower redshift. It is thus appropriate to start with a short history of high redshift quasars aligned with low redshift galaxies.

It has long been accepted that radio-emitting material is ejected, usually paired in opposite ejections, from active galaxies. The material is therefore aligned and points back to the galaxy of origin. But, and this is the major additional property of the associations, the ejected material frequently has a much higher redshift than the central galaxy. The prototypical pairs and alignments of higher redshift objects in this introductory section are taken from a body of data which is now too large to present completely. Nevertheless, it is hoped that the sample presented here will fix firmly the result that redshifts do not generally indicate recession velocity and are not reliable distance indicators. Even more importantly, the empirical data contained in these discordant associations is perhaps the only evidence capable of leading to a fundamental physical understanding of the origins of quasars and galaxies and the cause of intrinsic redshifts.

1

Quasars

A. Strong Pairs of Radio Quasars

It was a strong pair of radio sources, the famous 3C 273 and 3C 274 across the brightest galaxy in the Virgo Cluster (Arp 1967), which first indicated that a quasar could be at the same distance as a nearby, low-redshift galaxy, rather than at the much larger distance indicated by its redshift. In Fig. 1 here we show a pair of very bright Parks radio sources across the disturbed IC 1767 (Arp, *Astrofizika*, 1968). Both of these later turned out to be quasars of strikingly similar redshift.

Fig. 2 shows a pair of 3C radio quasars across the disturbed pair of galaxies NGC 470/NGC 474 (Arp, *Atlas of Peculiar Galaxies* No. 227). Quasars this radio bright are very rare (a total of 50 over the northern hemisphere). This yields a frequency of only one per 320 sq. deg., and a chance of only 5×10^{-6} of finding both so close to an arbitrary point in the sky. If we then calculate the chance that they are also accidentally aligned within

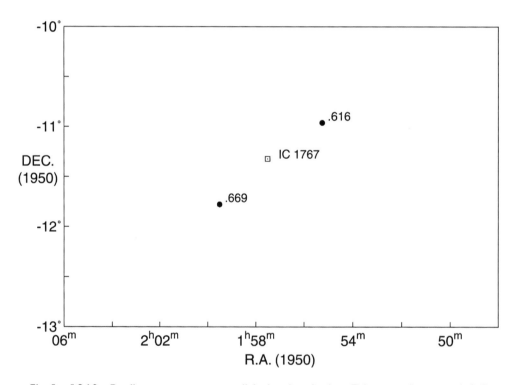

Fig.1 - 1968 - Radio quasars across disturbed galaxies. This very strong pair fell across a galaxy with z = .018 and later turned out have redshifts z = .616 and .669.

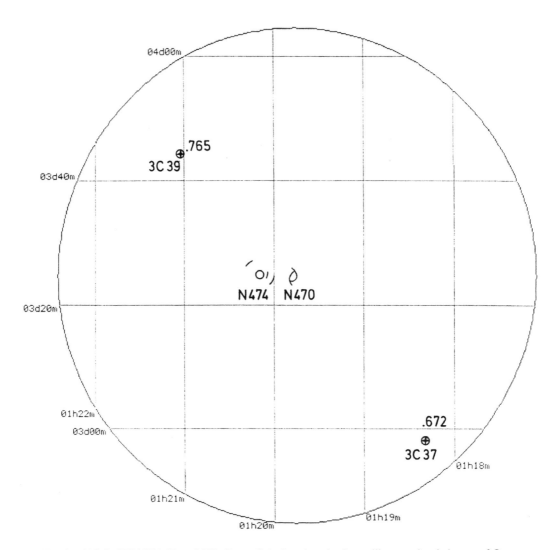

Fig.2 - NGC 470/474 (Arp 227). Two disturbed galaxies with a pair of strong, 3C radio quasars (z = .765 and .672) paired across them. Accidental probability of this alignment is 1×10^{-9}.

a degree or so, that they are equally spaced across the centroid within about 10%, and that their redshifts are within .09 of each other out of a range of about 2, then the probability that this might be an accidental association is about 1×10^{-9} (i.e., one chance in a billion). Note that this is not an a posteriori probability, because in the past 35 years many examples of paired quasars across active galaxies have been found with just these characteristics. In fact, it is confirmation of a predicted configuration at a significance level which should be considered conclusive.

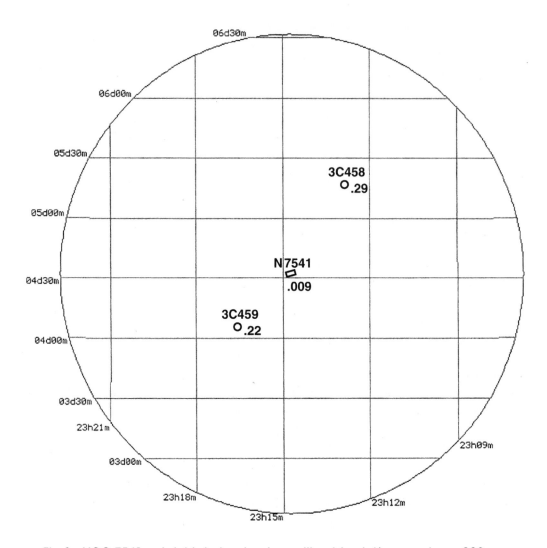

Fig.3 - NGC 7541, a bright starburst galaxy with a blue jet/arm and z = .009 falls between two strong, 3C radio sources which are now identified as quasars with z = .29 and .22.

An almost identical association is shown in Fig.3. There the central object is a bright starburst galaxy with a blue jet arm extending out to the WNW. Its redshift is z = .009, very close to the z = .008 of the galaxies in the preceding Fig. 2. Also as in the preceding example, two 3C radio quasars of very similar redshift (z = .22 and .29) are paired at only slightly greater distances across the active galaxy. Finding two such associations at this probability level is extraordinarily compelling. And, of course, Fig 1 shows a very similar pairing, except that it is in the southern hemisphere where the bright radio sources are from the Parks Survey rather than the 3C Cambridge Survey.

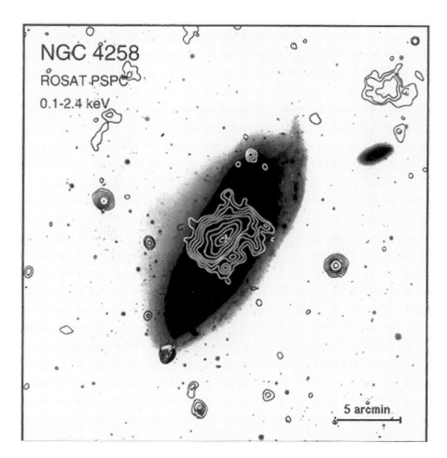

Fig.4 - 1994 -The famous active Seyfert NGC 4258 was found to have ejected two strong X-ray sources nearly along its minor axis. They turned out to be quasars of $z = .65$ and $.40$.

B. STRONG X-RAY PAIRS

When the X-ray satellites started reporting point sources that were frequently identified with blue stellar objects (BSO's), it quickly became clear that this was a much more certain way of discovering new quasars. As Fig. 4 shows, a strong pair of X-ray sources were observed closely across NGC 4258, a galaxy noted for its ejection activity. Before they were confirmed as quasars (E.M. Burbidge 1995), it was suggested that these X-ray sources had been ejected outward, closely along the minor axis of NGC 4258. (Pietsch et al. 1994).

Another very strong X-ray pair (119 and 268 cts/ks) was discovered bracketing the bright Seyfert galaxy NGC 4235. Fig. 5 reinforces an additional property of the quasar pairs, namely, that they tend to lie along the minor axis direction of the central galaxy. This seems logical in the sense

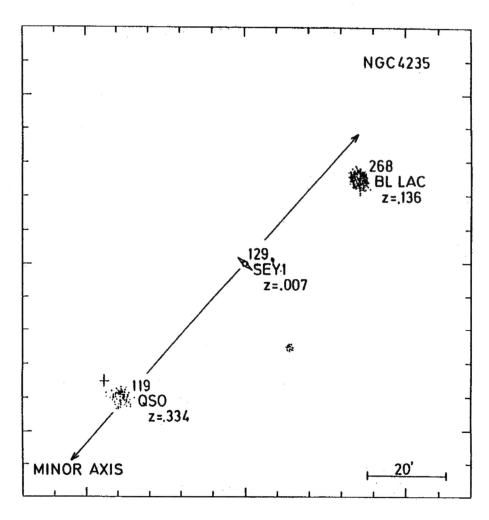

Fig.5 - 1997 - Two very strong X-ray sources (268 and 119 cts/ks) along the minor axis of an edge on Seyfert 1 galaxy. From a 7.5 sigma association of Seyferts with higher redshift quasars (Radecke 1997; Arp 1997).

that proto quasars would be able to exit the galaxy along the path of least resistance, i.e., the minor axis. The present *Catalogue* does not promote scientific theories. But as an aid in understanding the pictured relationships, it should be mentioned that in the variable-mass hypothesis (Narlikar and Arp 1993) the quasar starts as a low particle-mass plasmoid*, which would make it prone to interact with material at low latitudes in disk galaxies. Radio plasma, being more diffuse, would tend to be stripped away from a more compact X-ray emitting core through interaction with a galactic or intergalactic medium. Some observational evi-

---

* See Glosssary for definitions of plasmoid and other technical terms.

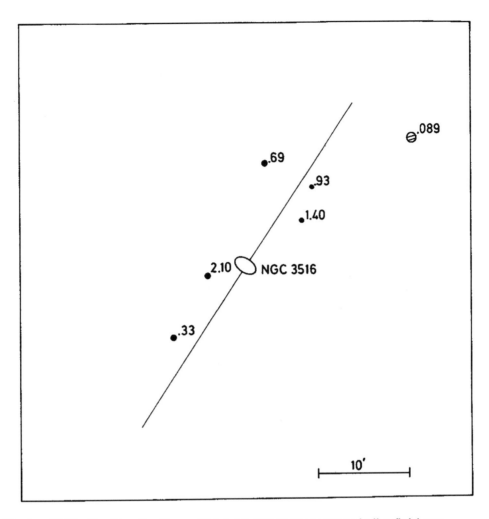

Fig 6. - 1998 - The Rosetta Stone. The brightest X-ray sources in the field are aligned along the minor axis in descending order of quantized redshift. The very active Seyfert has z = .009.

dence which may support this suggestion has been reported (Arp 2001). See also Appendix B at the end of this *Catalogue*.

c. Mutiple Quasar Alignments

Fig. 6 shows what I would nominate as the Rosetta Stone of quasar associations. Its validity is ensured by the fact that around this famous active Seyfert galaxy, NGC 3516 (Ulrich 1972), all the brightest X-ray sources have been identified optically and observed spectroscopically (Chu et al. 1998). They have been confirmed as quasars with the redshifts labeled in Fig. 6. It is readily apparent that they are distributed along a line drawn

NGC 5985

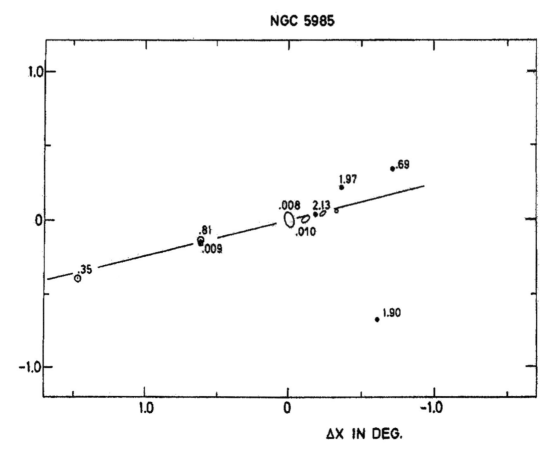

Fig.7 - 1999 - Five quasars falling along the minor axis of the bright Seyfert NGC 5985. Same descending order of quantized redshifts. Note low-redshift companions appearing along ejection line.

in the Figure. This line turns out to be the minor axis of NGC 3516. More-over they are ordered in redshift, with the highest redshifts falling closest to the galaxy and the lowest redshifts furthest away. The numerical value of each of the six quasar redshifts falls very close to one of the six most prominent quantized peaks of the Karlsson formula.

The next example of quasar alignments is almost as impressive, with five quasars aligned very accurately along the minor axis of the bright ap-parent magnitude Seyert galaxy NGC 5985. Fig. 7 shows that 4 of the 5 fall close to the redshift peaks, yielding, together with the previous asso-ciation, nine out of ten redshifts that obey the formula.

However, another aspect to the NGC 5985 alignment is very important for understanding the many regions presented in the following *Catalogue*.

My attention was drawn to this object by the fact that a low-redshift galaxy lay only 2.4 arcsec away from one of the quasars. It turns out that a total of four companion galaxies are found to lie along this same narrow quasar line. Their redshifts are only a few hundred km/sec higher than the redshift of NGC 5985 (Arp, *IAU Symp*. 194 p. 348-349), close enough to ensure they would be classed as physical companions, but with systematically higher redshifts, as has been found to be characteristic of groups dominated by large galaxies (Arp 1998a; 1998b).

The companions to spiral galaxies were long ago (Holmberg 1969) found to lie preferentially along the minor axis of the dominant galaxy, leading to the suggestion that the companions were formed in the parent galaxy and ejected outward. Their slightly higher redshifts would now imply that they were the end products of evolution from quasars where the intrinsic redshift component has decayed almost to zero.

Empirically, the observations of aligned companions at various redshifts are capable of explaining the persistent mystery of multiply interacting groups with discordant redshifts, such as Stephan's Quintet, and many other famous groups which often contain discordant redshifts. If the ejection direction stays relatively fixed in space, then ejecta in various stages of evolution and redshift can interact as they travel out along this line. But it is also possible—and probably observed here in the *Catalogue* examples—that material from the parent galaxy can be entrained along the supposed ejection path, thus furnishing another explanation of how low-redshift material could be found far from the central galaxy and aligned with higher redshift material.

One comment on why this picture has developed so slowly over the years is in order: The NGC 3516 paper was rejected by *Nature* magazine without being sent to a referee. Later it was demoted from the important short papers in *Astrophysical Journal Letters* to the not-so-pressing short papers in the main Journal. The data on systematic redshifts of companion galaxies was scarcely debated in the *Astrophysical Journal* main journal, and systematically rejected by referees and editors of the European Main Journal, *Astronomy and Astrophysics*.

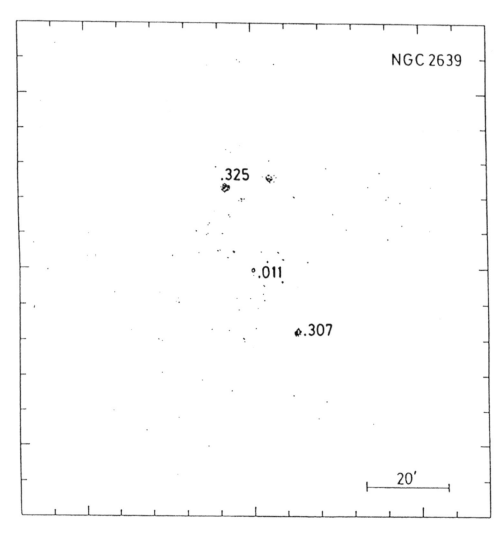

Fig. 8 - A pair of X-ray bright quasars across the Seyfert NGC 2639. Their redshifts differ by only .018.

D. ALIGNMENTS ROTATING WITH EPOCH

One example of an ejection axis that appears to have rotated with time is shown here in Fig. 8. The two brightest X-ray sources in the NGC 2639 field turn out to be quasars with redshifts differing by only .018 (Burbidge 1997). This would make it essentially impossible to argue that they were unrelated projections from the background. Their alignment is somewhat rotated from the minor axis of the central Seyfert. But closer to the galaxy, there is an extension of X-ray sources lying directly along the minor axis (Fig. 9).

10

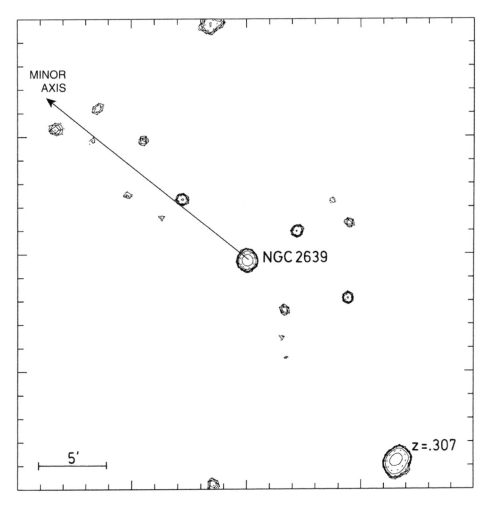

MINOR
AXIS

NGC 2639

5′

z = .307

Fig. 9 - Fainter X-ray sources, some identified as BSO's, lie precisely along the minor axis of NGC 2639 which has apparently rotated some degrees from the earlier ejection of the outer pair.

The only plausible explanation would seem to be that these sources represent a more recent ejection when the minor axis has rotated slightly from its earlier position. Obviously redshift measurements of the optical identifications in this more recent ejection would be invaluable aids to studying the evolution and interaction properties of these apparently younger quasars. The relatively short time allocations on moderate sized telescopes needed for these measurements have, however, not been forthcoming.*

---

* Four of the X-ray sources along the line in Fig. 9 have finally been confirmed as quasars with z from 0.337 to 2.63 by E.M. Burbidge in part of a night at the Keck 10 meter telescope in Hawaii.

11

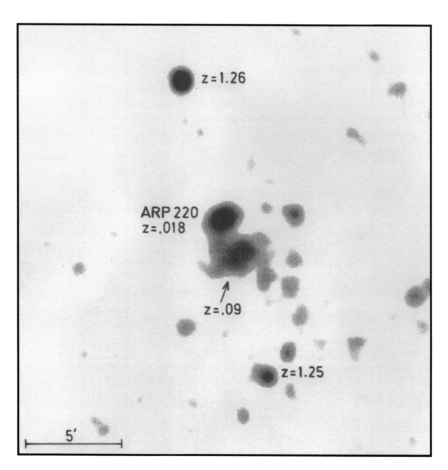

Fig. 10 - 2001 - X-ray quasars with almost identical spectra exactly aligned across the dust shrouded nucleus of the ULIRG, Arp 220. At the base of the trail of X-ray sources a group of z = .09 galaxies are connected by X-ray and radio material to the disturbed, infrared luminous galaxy (Arp 2001, Arp et al. 2001).

E. DISTURBED MORPHOLOGY OF ULTRALUMINOUS INFRA RED GALAXIES (ULIRG'S)

One of the supposedly most luminous nearby galaxies known is Arp 220 (Arp, *Atlas of Peculiar Galaxies* No. 220). The X-ray map in Fig. 10 shows that a pair of quasars across it have been found to differ by only .009 in redshift (Arp et al. 2001)! The pair is aligned as exactly as can be measured across this dust shrouded nucleus, whose activity is masked by an estimated 50 magnitudes of obscuration. If the two quasars have appreciable ejection velocity at the present time, then it is very unlikely that they are travelling so exactly across the line of sight that their radial velocity components would show negligible difference.

As suggested by previous cases, however, if the quasars have interacted strongly with the galaxy on their way out, they could have lost most of their ejection velocity. This explanation for the closely matching redshifts is attractive, because it also accounts for the extremely disrupted state of the ejecting galaxy. In turn, this may also be connected with the abundance of fine, solid-particle dust which results in the strong infrared radiation.

A second inference from the X-ray map in Fig. 10 arises from the fact that at the base of the southerly trail of X-ray sources which leads back to Arp 220 is a group of three or more galaxies with a redshift of $z = .090$. They are connected to the active galaxy by radio and X-ray bridges (Arp 2001). At their conventional redshift distance, these galaxies would be suspiciously bright, close to having the conventional luminosities of quasars, but they appear as elliptical and lenticular cluster galaxies. If they were younger, intrinsically redshifted ejecta which had been stopped, and then evolved close to Arp 220, they would instead represent higher redshift companion galaxies such as are found in many groups of galaxies. More discussion of this process can be found at the end of this *Catalogue* in Appendix B. But it is very important for the following *Catalogue* to introduce at this point empirical evidence that would support the cases where higher redshift *clusters of galaxies* appear physically associated with low-redshift, presumably ejecting, central galaxies.

F. PREFERRED VALUES OF REDSHIFT

Starting with Burbidge and Burbidge (1967), quasar redshifts in general were shown to occur in discrete values. Later Karlsson discovered that they obeyed the empirical law:

$$(1 + z_{i+1}) = 1.23(1 + z_i)$$

$$z = .06, \quad .30, \quad .60, \quad .96, \quad 1.41, \quad 1.96, \quad 2.64, \quad 3.48 \ldots$$

Quasar redshifts in many of the associations in the present *Catalogue* fall very close to these preferred values. Because it is unlikely that we are at the center of expanding shells, this would seem to require the dominant component of the redshift to be intrinsic. At some level, however, there should exist a component of peculiar velocity. Since the spread around the quantized values is observed to be of the order of $\Delta z = \pm 0.1$ (Arp et al. 1990), it is natural to suggest that the latter represents the (Doppler) velocity component of the redshift.

In the following section we actually compute the speed with which the quasars are moving through space. Indeed it turns out that the quasars

TABLE 1. WELL DEFINED PAIRS WITH REDSHIFTS. FIRST 7 OBJECTS ARE DISCUSSED IN ARP (1998) AND LAST THREE IN ARP (1998B).

Galaxy	z_G	z_1	(peak)	z_2	(peak)	vel_{ej}	vel_{ej}
NGC4258	0.002	0.653	(0.60)	0.398	(0.60)	0.031	−0.128
NGC4235	0.007	0.334	(0.30)	0.136	(0.30)	0.019	−0.132
NGC1068	0.0038	0.655	(0.60)	0.261	(0.30)	0.030	−0.034
NGC2639	0.0106	0.3232	(0.30)	0.3048	(0.30)	0.007	−0.007
IC1767	0.0175	0.669	(0.60)	0.616	(0.60)	0.025	−0.007
Mark205	0.070	0.464	(0.30)	0.633	(0.60)	0.052	−0.046
PG1211+143	0.085	1.28	(0.96)	1.02	(0.96)	0.072	−0.050
A/H\ #1	0.51	2.15	(0.96)	1.72	(0.96)	0.064	−0.081
A/H\ #2	0.54	2.12	(0.96)	1.61	(0.96)	0.034	−0.135
Her	0.55	2.14	(0.96)	1.84	(0.96)	0.034	−0.065

appear to be moving with respect to their parent galaxy with velocities from 10,000 km/sec at intrinsic redshift $z = 0.3$, to 30,000 km/sec at intrinsic redshift $z = .96$. The velocity of separation from the ejecting galaxy appears to fall as lower intrinsic redshifts are considered.

G. EJECTION VELOCITIES

To make this calculation we restrict ourselves to pairs of quasars. They are the most common association, and their approximately equal spacing across the central galaxy implies that they were ejected simultaneously with equal velocities. If momentum is conserved, one of the objects should have a component of velocity away from the observer, and the other toward the observer. We can test whether these expectations match observations.

Table 1 lists the best determined pairs of quasars lying across active galaxies for which redshifts have been measured. (There are more apparent pairs awaiting measurement.) The table lists the redshifts of the central galaxies (z_G) and the measured redshifts of the paired quasars (z_1, z_2). These observed quasar redshifts are then corrected to the galaxy center by means of $(1 + z_Q) = (1 + z_1)/(1 + z_G)$, and compared to the nearest

14

peak of quantized redshift given by the Karlsson series. The difference between z_Q and the nearest peak is assumed to represent the true ejected velocity of ejection v_{ej} (in units of c), i.e., $(1 + z_{ej}) = (1 + z_Q)/(1 + z_p)$. These values are listed in the vel_{ej} columns, 7 and 8, of Table 1.

As Table 1 shows, the only ambiguous cases are the low-redshift members of the NGC 4258 and NGC 4235 pair which would be closer to a peak if they were falling back in, rather than still moving outward. In all the other cases the z_Q's associated with the nearest peak denote one object moving away from the observer and one toward the observer. This result alone (8 out of 10) would confirm the hypothesis of ejection in opposite directions with the magnitude of velocities listed in Tables 1 and 2.

Prior confirmations of the peaks in redshift have been made on large samples. Nevertheless, it is impressive to see the conformity of the pairs listed in Table 1.

In addition to this quantization of the intrinsic redshifts, it is readily noticeable that the ejection velocities in the pairs are fairly well matched, with the away velocity about the same size as the toward component. This is very impressive because there are several factors which could cause a mismatch even if the quasars were ejected initially at the same instant with the same velocity. One factor is that the initial ejection might not be exactly in opposite directions. That would cause different projections of ejection velocities to be observed in the toward and away directions. More importantly, however, the quasars have to penetrate through different amounts of galactic and intergalactic medium in different directions. For pairs along the minor axis these considerations should be less important, although they could be involved in the few cases where the match is not as good as the average. Of course, the intrinsic redshifts must evolve downward *in steps* as the ejecta travel outward. Depending on how fast they make the transition from peak to peak, there will be a chance of catching some redshifts in transition between peaks.

An important quantity derived so far from the group of pairs in Table 1, then, is the average velocity of separation from the ejecting galaxy as a function of redshift. This is summarized in Table 2. In terms of an empirical model this means that the ejected quasar must have slowed down from its original velocity, assumed close to c, to about 28,000 km/s by the time the redshift has evolved into the $z = 0.96$ peak. After that it must slow to about 10,000 km/s by the time the $z = 0.30$ peak is reached. If the quasars are then to evolve into bound companion galaxies, they must essentially lose all their velocity by an apogee of about 500 kpc. Galaxy redshifts, although they have much smaller intrinsic components, have also been shown to fall at certain preferred values. The most conspicuous are the 72 km/sec Tifft quantization and the 37.5 km/sec Napier quantization. Another peak is apparent in the redshift of the Perseus-Pisces cluster, which can be seen all over the sky at $z \sim .017$. In the latter case, we are forced to ask whether these galaxies form an expanding shell with us at the center, or, alternatively, they are all at the same age and evolutionary stage distributed throughout a static volume at different distances from us.

TABLE 2. EJECTION VELOCITIES

Peak	Projected Δz		Average absolute value	Deprojected Average Velocity $(\times \sqrt{2})$ (kms^{-1})
0.30	0.026	–0.021	0.023	9,729
0.60	0.029	–0.060	0.045	19,077
0.96	0.051	–0.083	0.067	28,405

THE 3C SAMPLE OF ACTIVE GALAXIES AND QUASARS

The 3C Cambridge Survey lists between 400 and 500 of the brightest radio sources in the sky north of Dec. = –5 deg. Of those which are extragalactic, only 50 are quasars. The spectroscopic observation of the 3C sources was already complete enough by 1971 to show that these quasars fell closer to bright, low-redshift galaxies. For the whole sample of 3C sources the probability that this result was accidental is $<10^{-3}$ (Burbidge, Burbidge, Strittmatter & Solomon 1971—the famous B^2S^2 paper).

This result, however, was based solely on the criterion of nearness on the sky. In subsequent years some of the closest pairs have shown other evidence for association, and a number of additional high significance associations with 3C objects have been found. (Arp 1996; 1998b; 2001). If we ask what determines the probability of an association we can list five empirical criteria: nearness, alignment, centering, similarity of ejecta (usually z's or apparent mag.) and connections (bridges, jets and filaments). With these criteria, we can add at least 17 more associations of 3C quasars with low-redshift galaxies having chance probabilities ranging from 10^{-3} to 10^{-9}. This seems to take the case for physical association beyond sensible calculation. We briefly discuss three individual cases below because of the strong evidence they contribute for ejection and quantization.

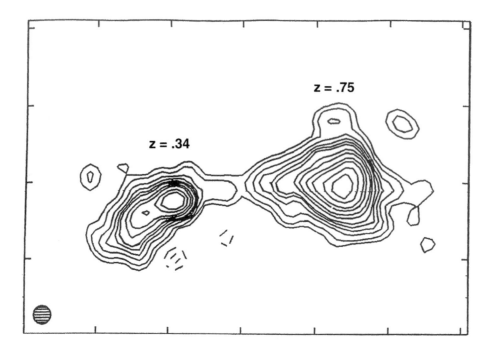

Fig. 11 2002 - Radio map at 1.6Ghz of 3C343.1 by Fanti et al. 1985. *Separation of sources is only 0.25 arcsec.* Note the opposite ejections from the radio galaxy, the western of which leads directly into the quasar. The compression of the radio contours on the west side of the quasar attests to its motion directly away from the galaxy.

A. 3C 343,1

In March 2002 Marshall Cohen called Margaret Burbidge's attention to a 3C radio source that had two redshifts. The abstract of the paper reporting this (Tran et al. 1998) ended with the statement: "Our data reveal a chance alignment of 3C 343.1 with a foreground galaxy, which dominates the observed optical flux from the system." It was a simple matter, however, to look up the high resolution radio map (Fanti et al. 1985) and find the two objects linked together by a radio bridge, as shown here in Fig. 11. We now calculate some probabilities of this being a chance alignment and show how the configuration follows the rules of many previous physical associations.

A circle of 0.25 arcsec radius subtends an area of 1.5×10^{-8} sq. deg. on the sky. In the now essentially completely identified 3C Catalogue there are about 50 radio quasars. Assuming 23,000 sq deg. to Dec. = −5 deg. we compute 2.2×10^{-3} such quasars per sq. deg., giving a probability of

3×10^{-11} of accidentally finding the $z = .750$ quasar within 0.25 arc sec of the $z = .344$ galaxy.

However, even if we do not consider the radio material linking them a bridge, we must still estimate the possibility that the radio tail from the galaxy points within a few degrees to the quasar and, similarly, from the quasar back to the galaxy. This would give a further improbability of $(\pm 2/90)^2 = 5 \times 10^{-4}$. *The combined probability of this configuration being chance is of the order of 10^{-14}.*

Because the galaxy and quasar together are faint, apparent mag. 20.7, it may be that the galaxy is fairly normal and lies at its considerable red-shift distance. That would also help explain the exceedingly small, apparent separation of the objects. There remains, however, an intriguing question about the numerical value of the redshift. The Karlsson preferred redshift values in this interval are shown below. (For references to the derivation of these peaks see Arp et al. 1990; Burbidge and Napier 2001 and also Section D in the description of the *Catalogue* which follows.) In this range the peaks are:

$$z = \ldots .30, .60, .96, 1.41, \ldots$$

The galaxy at $z = .344$ is close to the $z = .30$ value. But the quasar at $z = .750$ is about midway between the next preferred values. The solution to this apparent discrepancy is to compute the redshift of the quasar as seen from the rest frame of the galaxy:

$$(1 + z_0) = (1 + z_Q)/(1 + z_g) = 1.750/1.344 = 1.302$$

Hence the redshift of the quasar is $z = .302$, an almost perfect fit!

If the quasar were not physically associated with the galaxy this, of course, would be an additional improbable accident. This calculation is also important when samples of fainter quasars are considered, as noted in Arp et al. (1990). As for the distance of the $z = .344$ galaxy, it might, of course be closer than its redshift distance. Of interest in this connection is a pair of UGC galaxies a little over a degree away with $z = .032$ and .033. Moreover, in the direction of this pair from the $z = .344$ object are a quasar of $z = 1.49$ and the quasar 3C 343, with $z = .99$.

We have discussed this pair of objects from the standpoint of whether there could be any "a posteriori quality" to the extraordinarily small probability of a coincidental association. In fact, we have found that they were just more extreme values of the same properties that have characterized so many other physical associations of high significance—nearness, alignment, disturbances, connections. It is also striking to note

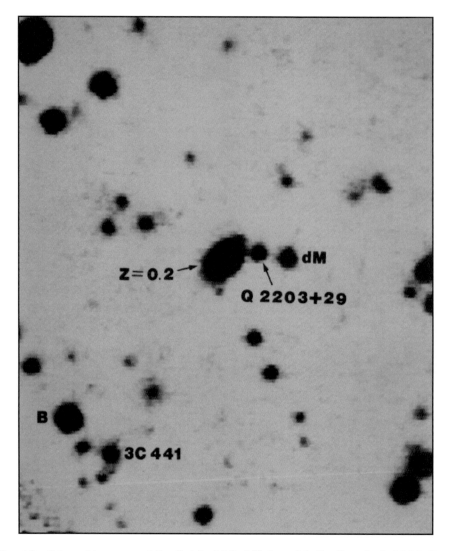

Fig. 12 - R band images of the field of 3C 441 from McCarthy et al. 1988. The galaxy at z = .202 appears either to have its west end occulted and/or a luminous connection to the quasar 2203+29, which has z = 4.399.

that this case has been circulating in the published literature for more than 4 years, and was even described as a foreground galaxy coincidentally close to a background quasar. One wonders how many other decisive pieces of information have gone unrecognized.

B. 3C3441

Fig.12 shows the area around 3C 441. About 36 arcsec from this 3C galaxy a quasar of z = 4.399 was accidently discovered (McCarthy et al.

1988). The chance of this being coincidental was estimated at a few times 10^{-3}. But what was ignored was that the quasar, at R = 20.8 mag., was either partially obscuring the end of, or exhibiting a luminous bridge to, a relatively bright galaxy at z = .202. (The image appeared at the limit of resolution, but apparently no effort was ever made to get a better picture.) What has been apparent for some time, however, is that the z = 4.399 redshift fits the Karlsson peak redshift in its vicinity fairly well... 1.96, 2.64, 3.48, 4.51... But in the rest frame of the z = .202 galaxy it fits almost perfectly at z = 3.49!

c. 3C 435

This 3C source turned out to be two sources about 12 arcsec apart, one of z = .461 and one of z = .865. The latter source appears to be exactly and indistinguishably superposed on a galactic star of about zero redshift (McCarthy et al. 1989). Apparently no attempt has been made to straighten out this intriguing situation by getting images with higher resolution telescopes. But putting that matter aside, it appears that the z = .865 object in the rest frame of the z = .461 object is z = .28—very close to the peak redshift value of z = .30.

D. A 2DF GALAXY

Not a 3C galaxy, but a composite spectrum reminiscent of the situations described above appeared recently among 55 quasars with z > .3 in the 2 degree *Field Galaxy Redshift Survey* (Madgwick, D. et al. 2001). The two spectra, appearing in a seemingly single object, have z = .1643 and z = .87. In the reference frame of the galaxy the quasar would go from .87 to .61!

"The Arp Controversy"

NGC 4319   z=0.006

Luminous bridge

Markarian 205   z=0.070

4x80 sec 50cm f/4 SX CCD   1998 March 20th   0010h U.T.   D.Strange   Worth Hill Observatory   Dorset U.K.

Fig. 13 (Plate 13 Intro) – The famous debate between big telescopes in the 1970's as to the reality of the connection between the galaxy NGC 4319 and the quasar/AGN Markarian 205 has been settled by these CCD frames taken by D. Strange with a 50cm telescope in the English countryside.

EJECTION ORIGIN OF QUASARS

Because radio sources and X-ray jets are believed to be ejected from galaxy nuclei, it was reasoned in the beginning of this Introduction that radio quasars and X-ray loud quasars were also ejected. The observed pairings in all these sources was strong support for this conclusion. The potential violation of redshift as a distance indicator, however, has caused opinion leaders in the field to demand ever more proof of the physical association of such discordant redshift objects. One form such proof could take is luminous connections from galaxies to higher redshift objects. There are a few cases where optical bridges and filaments are seen. One famous connection is between a low-redshift galaxy, NGC 4319 and an AGN/Quasar, Mrk 205. The progressive evidence from optical to X-ray was featured on the covers of my first two books, *Quasars, Redshifts and Controversies* and *Seeing Red: Redshifts, Cosmology and Academic Science* (Arp 1987; 1998b). The optical connection at least, once hotly debated by big telescope observers, appears to have been settled by an amateur in the English countryside with a 50 cm telescope (Fig. 13, and color Plate 13 Intro).[*] It is also interesting to note that the redshift of Mrk 205 ($z = .070$), when transformed to the rest frame of the disturbed galaxy ($z = .006$), becomes $z = .064$. This is very close to the first quantized redshift of $z = .06$.

We have seen evidence in radio contours for ejection of quasars. A particularly conclusive case was 3C 343.1 (preceding section). M 87 in the Virgo Cluster is a well-known case where a strong radio jet and enclosed inner X-ray jet, together, point along galaxy alignments to bright quasars. In other 3C objects such as 3C 275.1 and 3C232, X-ray jet/filaments have been found to point from the nearby active galaxy to the quasar (Arp 1996). But in the case of a very active object like M 82 (3C 281), it has been stated in numerous papers that X-rays are being ejected along the minor axis of this explosive galaxy. A very dense group of quasars is found in this direction immediately SE of M 82 (Arp 1999), and more BSO X-ray candidates are found NW, in the other minor axis direction, (G.R., E.M. Burbidge, H. Arp and Z. Zibetti, *ApJ* in press).

---

[*] Not yet! In Oct. 2002 a Hubble Space Telescope image (and press release) claimed no connetion. But many independent researchers printed the same picture and strikingly confirmed the bridge!! See cover picture and caption for further discussion.

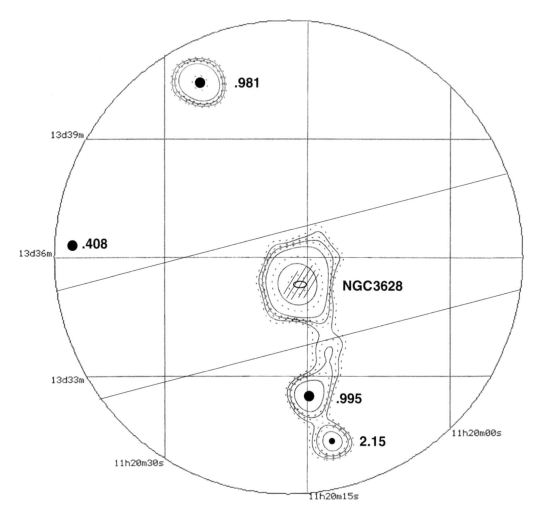

Fig. 14 - Quasars of z = .981 and .995 are paired across the minor axis of the bright, low-redshift galaxy NGC 3628. The isophotes show that an X-ray jet from the center of the galaxy contains the z = .995 quasar and ends on the z = 2.15 quasar. (see Arp et al. 2002 for other features of this association.)

A recent case where X-ray observers have identified ejection along the minor axis, however, is the bright galaxy in the Leo triplet, NGC 3628. There neutral hydrogen flow out of the galaxy is observed, as well as X-ray ejection. An excess of quasars detected by objective prism observations as well as X-ray quasars has been discovered in this ejection along the minor axis of NGC 3628. X-ray quasars of z = .981 and .995 are paired across the galaxy and aligned along this minor axis. *But the most decisive observation establishing ejection is the fact that two of the quasars along the minor axis fall exactly in, and at the terminus of, an X-ray jet emerging from the nucleus of NGC 3628.* They have been, so to speak, caught in the act of departure! One Figure is shown here (Fig.14), but the published paper contains diagrams and pictures of the various kinds of material outflowing from the galaxy (Arp et al. 2002). This should unequivocably settle the fact that quasars are ejected from galaxies.

A. GAMMA RAYS—THE MOST ENERGETIC CONNECTION

In 1995 an X-ray survey map of the Virgo Cluster was published. It showed X-ray connections between the dominant galaxy in the center (M49) and the radio galaxy 3C 274 to the north and the quasar 3C 273 (z = .158) to the south. This result was denied publication in major journals and ignored. But then Hans-Dieter Radecke (1997) courageously published the gamma ray map, confirming the connections with ≥100 MeV photons, *except that the bridge was now much stronger in the southern connection to 3C 273, and continued on unmistakably to join the quasar 3C279 (z = .538).* The extremely high energy radiation was interpreted as the ejection of proto quasar material from the active nuclei of the older galaxies (Arp, Narlikar and Radecke 1997). The original Radecke map can also be seen in *Seeing Red* Plate 5-18 where more details of the events connected with it can be consulted.

In my opinion Radecke's gamma ray map of the Virgo Cluster is one of the most important and unequivocal findings in the subject of the distances, nature and origin of quasars. Yet it has been deliberately ignored, and Radecke himself is no longer involved in professional research.

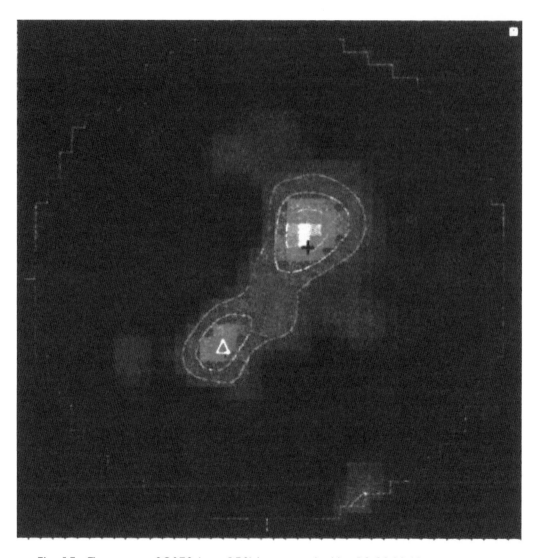

Fig. 15 - The quasar 3C273 (z = .158) is connected by 10-30 MeV gamma rays to the quasar 3C279 (z = .538). This is the latest COMPTEL map as published in "Research 2000-2001, a book of posters," Max-Planck-Institut für Extraterrestrische Physik. (See Radecke 1997b; Arp et al. 1997; 1998b for even higher energy maps.)

Perhaps the most important confirmation of this discovery is the fact that slightly lower energy gamma rays, 10-30 MeV, showed the same unmistakable connection between the quasars of z = .158 and z = .538! This latter result, shown here in Fig. 15, was obtained with a completely different instrument, the Compton scattering COMPTEL, as opposed to the photon counting EGRET. Even with further observations added, the lower en-

26

ergy gamma rays confirmed the highest energy connection. The absolutely crucial question for the profession of science then is posed: Why is this high-energy photon map, which overturns the most basic assumption in extragalactic astronomy, suppressed and ignored?

INTRINSIC REDSHIFTS OF GALAXIES

The evidence from the associations in the *Catalogue* can best be interpreted as high-redshift quasars evolving to lower redshift, then into active galaxies and finally into normal, low-redshift galaxies. It is instructive, therefore, to study the intermediate phases in this evolution by examining physical companions of variously higher redshifts associated with low-redshift parent galaxies. It has been found that the former tend to be higher surface brightness, active, non-equilibrium forms. This supports their classification as the next stage of the compact, energy dense quasars. Of course, establishing even one or two cases of galaxies which have clearly non-velocity redshifts raises the question of a physical mechanism that can account for such a redshift. Moreover, a non-velocity redshift of an extended, well resolved companion rules out mechanisms which are frequently proposed, such as gravitational redshift or tired light. This is because the light travels essentially the same path to us from both the high and low-redshift galaxy—and, further, all parts within the high-redshift galaxy—the stars, gas, dust etc.—are redshifted about equally.

In the time taken to evolve from quasar to normal galaxy the objects can drift from their original ejection patterns. The average cone angle for companions around the minor axis of parent galaxies is ±35 deg., as opposed to ±20 deg. for quasars (see "The Origin of Companion Galaxies," Arp 1998a.) As a result, the identification of excess redshift companions becomes more of a statistical calculation based on their nearness or grouping around the parent galaxy. On the other hand, if distances independent of redshift can be established for such galaxies, they can be compared directly to their redshift distances. These cases, as well as interaction evidence, can establish individually the presence of non-velocity redshifts.

The observational evidence on discordant redshifts in groups is voluminous and goes back to 1961 when Geoffrey and Margaret Burbidge took spectra of the components of Stephan's Quintet (see e.g., Arp 1987; 1998b). But here I would like to start with recent results on the distances of galaxies and trace their connection to some of the highlights of past distance discrepancies.

CEPHEID DISTANCES AND THE HUBBLE CONSTANT

In the early 1950's the Period-Luminosity law of Cepheid variable stars was calibrated in open clusters in our galaxy. It was used to obtain distances to galaxies in neighboring groups to our own. Dividing the redshifts of those galaxies by their distances yielded a Hubble constant near $H_0 = 50$ km/sec/Mpc. There was always controversy over this value, however, with some investigators getting larger values. Since this constant was supposed to represent the expansion velocity of our universe, the larger values lead to an expansion age of our universe that was younger than or uncomfortably close to the age of the oldest stars in our galaxy.

In an effort to minimize the effect of possible peculiar (non systematic expansion) velocities on the determination of H_0, fainter Cepheids in higher redshift galaxies were measured with Hubble Space Telescope (HST). But Fig. 16 here shows that the $H_0 = 72 \pm 8$ which was officially celebrated, actually means serious trouble. The reason for this is that the majority of points define a nice, low dispersion line at about $H_0 = 55$. This is in keeping with Sandage/Tamman estimates of a Hubble flow which is quiet to about 50 km/sec dispersion in velocity, and also in consonance with quantization of galaxy redshifts at 37.5 km/sec which would be washed out with larger peculiar velocities.

But more distant than about 15 Mpc the relation explodes! The peculiar velocities jump to 1,000 km/sec and *become overwhelmingly positive*! The Hubble constant, used in standard cosmology to measure the expansion age of the universe, is clearly indeterminate. Since, however, discrepancies between expansion ages and oldest star ages have now been overridden by stepping on the "dark energy" gas pedal or applying the "dark matter" brakes, the value of H_0 has therefore become irrelevant in conventional cosmology.

Could these discrepancies be velocity caused? The answer is no on four counts:

1. The low velocity dispersion in the rather large ($r \leq 15$ Mpc) local neighborhood should not suddenly increase by an unacceptable amount.

2. Peculiar velocities should not be predominantly positive.

3. Tully-Fisher distances for much larger samples of galaxies all over the sky show the same effect.

REDSHIFT-DISTANCE HUBBLE DIAGRAM

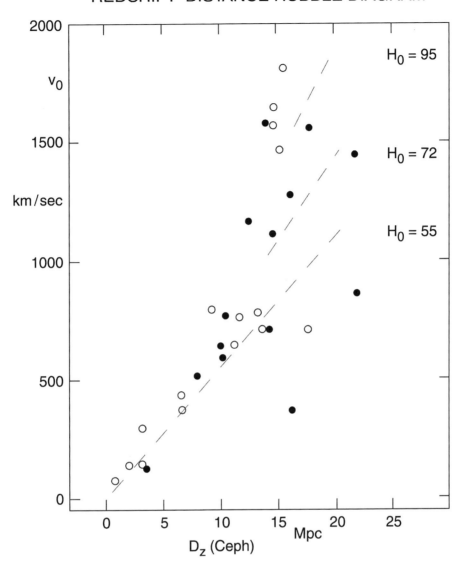

Fig. 16 - The redshift-distance plot that defines the Hubble constant. The distances are from HST measures of Cepheid variables and the redshifts are catalogued, galactocentric values (v_0) from Sandage and Tamman (1981). (See Arp 2002 for original paper.)

4. Evidence from associations of galaxies has been showing intrinsic redshifts for these same kinds of galaxies for over 30 years.

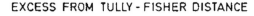

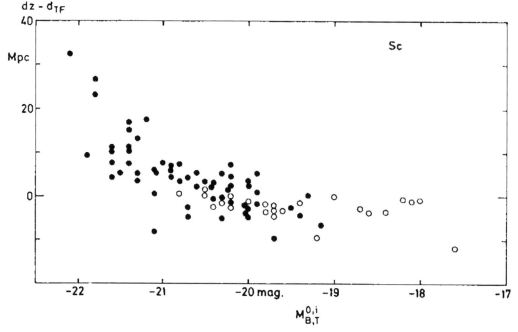

Fig. 17 - Absolute blue magnitudes plotted against the redshift distances minus Tully-Fisher distances ($d_z - d_{TF}$). Redshift distances for Sc galaxies are much greater than Tully-Fisher distances for Sc's with redshifts $z \geq 1000$ km/sec (filled circles).

A. TULLY-FISHER DISTANCES

As is well known, the only major alternative to Cepheid or bright star distances to galaxies is to measure their rotational velocities, infer their mass and thus luminosity, and then use the difference between their apparent and absolute magnitudes to calculate their distance. Fig. 17 shows the same result as Fig. 16, namely, that the galaxies of less than redshift about 1000 km/sec are well behaved with the Tully-Fisher (d_{TF}) distance, giving closely the same distance as the redshift distance (d_z). Above redshifts of 1000 km/sec the redshift distances become vastly greater than the TF distances. *But these are the same kind of galaxies that violate the redshift-distance relation in Fig. 16.*

One of the most active researchers in this field, David Russell, has checked the rotational distance criteria with another distance criterion— diameter as a function of morphological type. He finds very close support for the TF distances. But when he uses the redshift distances for these

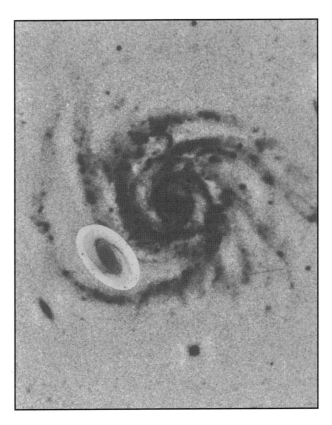

Fig. 18 - At its supposed redshift distance, the ScI galaxy NGC 309 is compared to M81, one of the largest galaxies of whose dimensions we can be sure. NGC 309 would be so enormous that it would make M81 look like a knot in one of its spiral arms.

same galaxies, he gets unprecedentedly large diameters. These are all luminosity class ScI-II galaxies (ScI being a classification based on strong, well defined spiral arms which I would identify as recently formed in ejection events and therefore generally younger galaxies.) The luminosity class I galaxies are the ones that deviate the most from the average line in Figs. 16 and 17. What this is telling us is that there is something wrong with the redshifts of these kinds of galaxies. They must contain a large intrinsic component!

B. WHAT DO THESE GALAXIES LOOK LIKE?

As a last resort, some astronomers might actually look at the objects they are using to calculate numbers. An example is NGC 309. It is an ScI, which is shown in Fig. 18. At its redshift distance, it is compared to a gal-

31

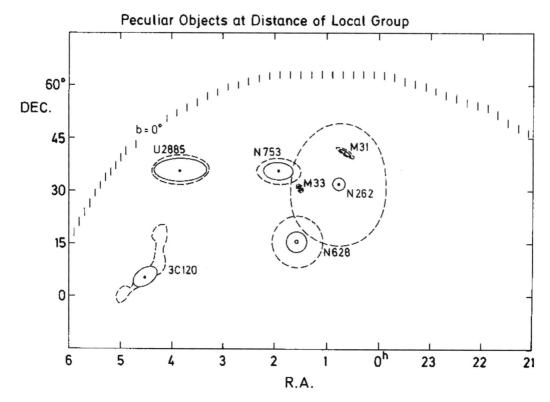

Peculiar Objects at Distance of Local Group

Fig. 19 - Galaxies whose redshift distances are the most excessive happen to fall in the direction of our Local Group. They are shown with optical boundaries as a solid line and Hydrogen by a dashed line. They are so big at their supposed redshift distances that they would fill the whole Local Group (from Bertola et al. 1998).

...axy to which we really know the distance. We see that the giant in our neighborhood, known as M 81, is swallowed like a knot in the arm of this supposedly monstrous NGC 309. This revelation usually shocks astronomers, because they never think about how they casually accept objects which contradict their empirical picture, but for which there is no precedence or independent observational support.

In fact NGC 309 looks rather like an ordinary spiral, of which there are many examples that are fainter and smaller than galaxies like M 81 and M 31. If NGC 309 was really as large as its redshift distance would have us believe, it should furnish a supernova about every three years—a frequency amateur supernovae observers could easily testify is not seen.

Are there other examples of galaxies that would have improbable dimensions if situated at their redshift distances? Fig. 19 shows an array of galaxies, three Scl spirals: UGC 2885, NGC 753 (33 Mpc more distant based on its redshift than its 47 Mpc TF distance) and the bright apparent magnitude spiral NGC 628. If it were at its redshift distance, however, NGC 262 (Mrk 348) would have a diameter large enough to encompass the whole center of the Local Group. It would subtend more than 30 deg. on the sky! Yet it appears to be a dwarf spiral and has hardly more than 100 km/sec internal redshift differences. If the objects in Fig. 19 were really the sizes given by their conventional redshift distances, they would produce from 5 to 50 supernovae a year! As one might guess from their NGC numbers, these galaxies are actually located at the pictured positions in the sky. They could be high intrinsic redshift members of the Local Group. There are many groups in the following *Catalogue* which contain even larger ranges of redshift.

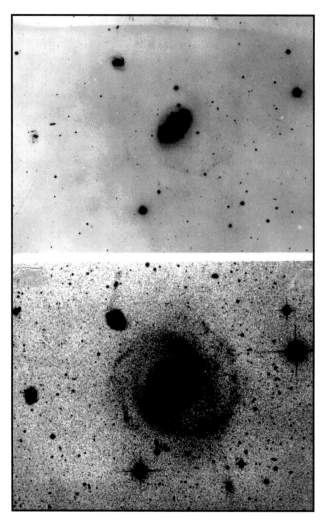

Fig. 20 - The Scl galaxy NGC 4156 (cz = 6,700 km/sec) is only about 5
arcmin NE of the bright Seyfert NGC 4151 (z = 964 km/sec). A deep photograph with the 200-inch at Palomar shows outer spiral arms leading toward
the NGC 4156 at +5,700 km/sec excess redshift, and also to a companion
SW at + 5,400 km/sec. (See also Fig. 18 in Appendix B, Arp 1988b and Arp
1977 for original paper.)

C. COMPANION GALAXIES AND LATE TYPE SPIRALS

As early as 1970, when *Nature* magazine was still publishing observational tests of astronomical assumptions, data appeared which showed
that companion galaxies and spirals had sytematically excess redshifts
relative to earlier type galaxies (Arp 1970; Jaakkola 1971; Arp 1990). Fig.
20 above shows a quintessential Scl galaxy, NGC 4156, which has been

known for decades to have a large excess redshift at the distance of its parent Seyfert Galaxy. Both are X-ray sources, while the companion at the end of the opposite arm is also an X-ray source and has the same large, excess redshift as NGC 4156.

Sc galaxies in particular showed this effect, and the most extreme form—ScI's with sharply defined arms—showed it the most conspicuously. The latter, including such ScI's as NGC 309 and NGC 753, have the greatest excess redshifts over HST Cepheid and TF distances, as shown in Figs. 16 and 17. There are even more extreme cases, as shown in Fig. 20, an ScI of cz = 6,700 km/sec on the end of the arm from NGC 4151 (an Sb with cz = 964 km/sec). (See even deeper images of NGC 4151 in Fig. 18 in Appendix B.)

However, the diffuse, low surface brightness X-ray galaxy on the SW arm shows that material with this intrinsic redshift can occasionally be disrupted and spread out into a non-spiral, non ScI form.

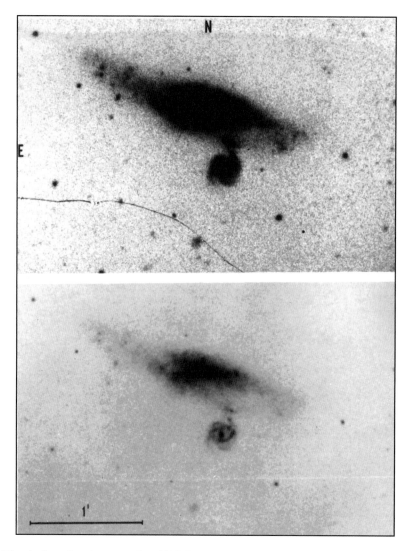

Fig. 21 - A dwarf galaxy north of NGC 4151 which has cz = 1,045 km/sec, with a ScI spiral of cz = 29,400 km/sec attached. (See Arp 1977 for original paper).

Here Fig. 21 shows another ScI of cz = 29,400 km/sec connected to a nearby galaxy of cz = 1,045 km/sec. In a short span one observer found 38 more examples of excess redshift companions reaching up to Δcz = 36,000 km/sec (Arp 1982). Unfortunately the latter discoveries have been consistently ignored in the last two decades.

As mentioned above, with the advent of X-ray and gamma ray observations, hard energy jets and connections were found from low-redshift galaxies to high-redshift quasars and ejected companions. The care with which these observations were reduced and presented to the astronomical readership can be judged by following some of the references in papers appearing since the 1970's. (Arp, 1996; Arp, Narlikar and Radecke, 1997; Arp, Burbidge, Chu, Flesch, Patat and Ruprecht, 2002).

In the following *Catalogue* companion galaxies and quasars are seen strung out over generally larger arcs in the sky. Gradations and similarities of redshift usually form the evidence for physical association with central galaxies. It may be that deep, wide-field imaging can furnish further connective evidence for association. But it may also be profitable to look close to the central galaxies with deep imaging in various wavelengths to capture associations in earlier stages. It is perhaps serendipitous that so many large telescopes and advanced detectors have already been built, which could be eventually be used to investigate a more sophisticated and complete physics.

CROSSING THE BRIDGE TO A MORE CORRECT PHYSICS

It is now 30 years since Fred Hoyle gave his Henry Norris Russell Prize Lecture before the American Astronomical Society in Seattle. Earlier in the meeting I had given a short resumé of the evidence for birth of quasars and their evolution into galaxies. It was therefore a thrill for me to later listen to Sir Fred outline a broad and insightful analysis of the kind of physics that we would need in order to deal with these observations in extragalactic realms. I was surprised to hear him offer, as a proof for the need to consider fundamental particle masses evolving from zero, my observations of the 16,000 km/sec companion attached by a filament to the 8,000 km/sec Seyfert galaxy NGC 7603. (The latest exciting news on this object is discussed at the at the end of Appendix B.) Five years later his former student, Jayant Narlikar, made a more general solution of the field equations than the Friedmann solution, which had launched the Big Bang in 1922 (see Narlikar 1977; Narlikar and Arp 1993; Arp 1998b.). The newer solution with evolving particle masses, I believe, elegantly explains the redshift associations which are still so disturbing to cosmologists and physicists.

But the subsequent story of what happened to Fred's lecture illuminates the situation of cosmology today. One leading astronomer came up to us as we were talking after the lecture and blurted out, "You are both crazy." His prestigious Russell lecture, which was traditionally published in the *Astrophysical Journal*, was inexplicably sent to a referee. Fred was outraged (as were others when they heard about it) and refused to proceed with publication. I endeavoured to convince him that it should be published, and he agreed to let me publish it in the book called *The Redshift Controversy* (ed. George Field, 1973). This book records the debate between myself and John Bahcall held at the American Association for Advancement of Science in Washington on December 30, 1972. If I had not been able to include it in this book, this seminal path to the future mapped out by one of the most eminent scientists of this era would have never even been available in the recorded literature.

In that lecture, entitled "The Developing Crisis in Astronomy," he ended by saying "...[the observations are] forcing us, whether we like it or not, across this exceedingly important bridge [to a more fundamentally correct physics]..." I now personally regret that a generation has passed and we are further than ever from making that advance. I hope that the following *Catalogue* of extragalactic objects will direct our feet back onto that bridge to a better future.

REFERENCES

Arp, H. 1967, ApJ 148, 321

Arp, H. 1968, Astrofizika (Armenian Acad. Sci.) 4, 49

Arp, H. 1970, Nature 225, 1033

Arp, H. 1977, ApJ 218, 70

Arp, H. 1982, ApJ 263, 54

Arp, H. 1987, Quasars, Redshifts and Controversies (Interstellar Media, Berkeley)

Arp, H. 1990, Astrophys. and Space Science 167, 183

Arp, H. 1996, A&A 316, 57

Arp, H. 1997, A&A 328, L17

Arp, H. 1998a, ApJ 496, 661

Arp, H. 1998b, Seeing Red: Redshifts, Cosmology and Academic Science (Apeiron, Montreal)

Arp, H. 1999, ApJ 525, 594

Arp, H. 2001, ApJ 549, 780

Arp, H. 2002, ApJ 571, 615

Arp, H., Bi, H.G., Chu, Y., Zhu, X. 1990, Astron. Astrophys. 239, 33

Arp, H., Burbidge, E.M., Chu, Y. Zhu, X. 2001, ApJ 553, L11

Arp, H. Burbidge, E.M., Chu, Y, Flesch, E., Patat, F., Rupprecht, G. 2002, A&A 391, 833..

Arp, H, Narlikar, J., Radecke, H.-D. 1997, Astroparticle Physics 6, 387

Bertola, F., Sulentic, J. Madore, B. 1988, New Ideas in Astronomy, Cambridge University Press

Burbidge, G.R., Burbidge. E.M. 1967, ApJ 148, L107

Burbidge, G.R., Burbidge. E.M., Solomon, P.M., Strittmatter, P.A. 1971, ApJ 170, 233

Burbidge. E.M. 1995, A&A 298, L1

Burbidge, E.M. 1997, ApJ 484, L99

Burbidge, G.R., Napier, W. 2001, AJ 121, 21

Chu, Y., Wei, J., Hu J., Zhu, X., Arp, H. 1998, ApJ 500, 596

Fanti, C., Fanti, R., Parma, P., Schilizzi, R., van Breugel, W. 1985, A&A 143, 292

Field, G., Arp, H., Bahcall, J. 1973, The Redshift Controversy, W.A. Benjamin Co., reading, Mass.

Holmberg, E. 1969, Ark. Astron. 5, 305

Jaakkola, T. 1971, Nature 234, 534

Madgwick, D., Hewett, P., Mortlock, D., Lahav, O. 2001, astro-ph/02033307

McCarthy, P., Dickinson, M., Filippenko, A., Spinrad, H., van Breugel, J. 1988, ApJ 328, L29

McCarthy, P., van Breugel, J., Spinrad, H. 1989, AJ 97, 36

Narlikar, J. 1977, Ann. Physics 107, 325

Narlikar, J. Arp, H. 1993, ApJ 405, 51

Pietsch, W., Vogler, A., Kahabka, P., Jain A., Klein, U. 1994, A&A 284, 386

Radecke, H.-D. 1997a, A&A 319, 18

Radecke, H.-D. 1997b, Astrophys. Space Sci. 249, 303

Sandage, A., Tamman, G. 1981. A Revised Shapley Ames Catalogue of Bright Galaxies, Carnegie Institution of Washington.

Tran, H., Cohen, M., Ogle, P., Goodrich, R., di Serego Alighieri 1998, ApJ 500, 660

Ulrich, M.-H. 1972, ApJ 174, 483

The Catalogue

ABOUT THE CATALOGUE

The *Catalogue* is a picture book that shows distributions of extragalactic objects in various sized regions of the sky. The maps presented here depict associations of quasars, galaxies, clusters of galaxies and related objects in patterns which are characteristically repeated. I believe it is possible for non-specialists and even specialists to simply glance at this succession of maps to understand the essential principle involved—alignments of higher redshift objects originating from larger, usually active, galaxies of lower redshift. If desired, from there it is a matter of each individual's judgement to consider models of ejection and evolution, with their consequences for physical processes such as the nature of redshifts, mass, time, gravity and cosmology.

The second purpose of these pictures is to furnish key objects for further observations. In almost all cases more redshifts, direct images, X-ray and IR observations would further test the validity or non-validity of the associations and also furnish important new information on their nature and origins. Having independent observers confirm new evidence on the processes which give rise to these patterns is perhaps the only way in which the majority of scientists will be led to accept the new paradigm they represent.

I hesitate to call this a *Catalogue* because it is not complete. Indeed, wherever I look in the sky—for example to discover where a certain active galaxy cluster, quasar or proposed gravitational lens came from—I am likely to find its source plus other families of extragalactic objects, with a large, low-redshift galaxy and associations of higher redshift companions. There are many more examples of this basic pattern to be discovered, so this is merely a sample. And again, their acceptance will be hastened by independent discovery.

ABOUT THE LISTED ASSOCIATIONS

The examples are listed in order of increasing right ascension (R.A. epoch 2000). This is to facilitate observers finding the kind of association they want to study in a region of the sky accessible to them. Each association is generally named according to the brightest object or the object considered to be the origin of the higher redshift objects. The picture is displayed on the right hand page with the most important information labelled. The title object is described at the top of the left hand page, its apparent magnitude and redshift if known, and any indications of an energetic nucleus, for example: Seyfert, infrared, X-ray activity or morphological distortion.

Next are listed the most significant objects for the association, usually bright or unusual, higher redshift objects which are aligned. Such objects are usually rare and completely surveyed, or readily assigned a uniform limiting magnitude. Since the brightest objects in a given class are the nearest to us, their associations stand out most conspicuously against background objects. Thus each field represents the starting information to launch a more complete investigation; for example whether there are somewhat fainter or different classes of objects in the field which reinforce the connection with the proposed object of origin. Many of these further observations can be carried out with small to medium aperture telescopes. This opens a critically important field of investigation to astronomers who do not have access to large telescopes and to amateurs who have modern CCD detectors and spectroscopic capabilities.

SUGGESTED USE

Further investigations are most easily initiated with the aid of computer archives. The currently available data for any plotted object in a field can be obtained from SIMBAD or NED. The R.A., Dec. and apparent magnitude can be used to target any galaxies that require redshifts. Infrared catalogues can be used to study IRAS sources. Catalogues such as NVSS and FIRST can be used to locate radio sources. For X-ray sources, the ROSAT archives from Max-Planck-Extraterrestriche (MPE) give X-ray measures in all-sky, PSPC and high resolution (HRI) modes (both sources and browser, the latter of which gives standard reduction maps under the click marked b). The X-ray sources are particularly important because point X-ray sources are often associated with blue stellar objects (BSO's). These, in turn, invariably turn out to be quasars that are easily identifiable spectroscopically. Optical identifications can be made through the automatic plate measuring surveys of blue and red Schmidt Sky Surveys (APM) and U.S. Naval Observatory (USNOA). Finding charts can be downloaded from the ESO digitized sky survey. The latter are useful for identifying bright extended X-ray sources (EXSS) as galaxy clusters and obtaining a first impression of their shapes.

No attempt has been made in the current sample of associations to list published references or their authors except in rare cases where previous studies might not be listed in the abstract data services (ADS). Any objects identified in either SIMBAD or NED will have references appended if they appear in modern, standard journals.

One must be wary of selection effects. In any given field the recording of any given type of object may be incomplete. For example, galaxy surveys might have a border passing through the field. Quasars could be sampled in a small, deep field or within an approximately 1 degree radius PSPC field.

But for objects like NGC galaxies, Abell galaxy clusters (ACO), 3C radio sources and bright apparent magnitude quasars, I assume essentially uniform and complete sky coverage. When it comes to fainter objects, e.g., 16th mag. galaxies or emission line objects, we might see clusters, but immediately ask if galaxies in the surrounding regions have been completely observed. If we can see the cluster elongated, however, we tend to believe that to be real, because there is no reason for cataloguers to measure along a line and no reason for them to be aware that the line pointed to a nearby, low-redshift galaxy, as occurs in a number of cases. The challenge that immediately presents itself is to observe more objects along this line—to test, for example, whether objects of similar redshifts support the conclusion that it is neither accidental nor a background feature. As observations on any candidate association are completed new objects will need to be investigated, and if they add to the understanding of the origin of the association they will represent important new, first hand discoveries.

WHAT TO LOOK FOR

A. PAIRS AND ALIGNMENTS

A sample of some of the best cases of quasars aligned across active, ejecting galaxies is discussed in the Introduction (prototypical aligned pairs). The purpose of that section was to establish firmly the pattern of high-redshift objects paired across a low-redshift, usually bright galaxy. It also established that the high-redshift objects tend to come out along the minor axis (when that is measurable), that they tend to resemble each other in redshift and other properties, and that redshifts tend to fall near preferred, quantized values (Arp 1998).

43

When young objects leave the galaxy of origin unimpeded, say along the minor axis, the the lowest redshift objects tend to be observed most distant, and higher redshift objects closer to the galaxy of origin. This is expected if the intrinsic redshift diminishes as a function of age, because the more distant ones would have been travelling longer. On the other hand, when they exit through an appreciable part of the galaxy, they tend to disrupt the galaxy, and the ejecta tend to fragment and stay closer to the galaxy of origin (see e.g., Arp 1999). Moreover, remnants of the original galaxy which have the same redshift as or slightly larger redshifts than the ejecting galaxy can be entrained along the ejection path leading to much higher redshift ejecta.

B. NEW EVIDENCE FROM ELONGATIONS OF GROUPS OF X-RAY SOURCES

As the data on the present associations was being collected, X-ray archives revealed that many of the apparently ejected higher redshift objects had been observed as X-ray sources. Simply looking at the plots in the ROSAT X-ray source or sequence browsers frequently showed that the recorded sources in the field were conspicuously distributed in elongated patterns, often across a central source. The fact that in a number of cases they were aligned toward a central object of origin would seem to be decisive proof that they originated in the (usually lower redshift) object.

In one key case a high resolution X-ray observation showed a cold front (in the nature of a bow shock) moving down the elongated X-ray cluster Abell 3667 at 1400 km/sec directly away from the central, X-ray ejecting, lower redshift galaxy. (This association is not presented in the main body of the present *Catalogue*; but see Appendix B here, and full details are given in Arp 2001 + note added in manuscript.) All the cases in the present *Catalogue* should be checked for X-ray properties. Since the energetic X-ray activity tends to decay, it should be indicative of younger objects and hence higher redshifts. The configurations also apparently can be relatively unrelaxed and therefore record the directions and processes involved in their origins. Chandra and XMM observations would be particularly valuable in this respect.

C. EVOLUTION

One of the most valuable types of data to be obtained from the present *Catalogue* associations is the behaviour of the ejected objects as they progress further from their galaxy of origin. Generally they reduce in redshift, presumably as they grow older and evolve. The morphological and energetic continuity as they transform from compact high energy density quasars to more quiescent, relaxed normal galaxies is very important in justifying the conclusion that quasars are continually evolving into normal galaxies. It also represents the best opportunity to obtain data on what fundamental physical processes are taking place. The associations presented in this *Catalogue* will, I hope, furnish the empirical data with which to trace and understand this evolution. (We should also be aware, however, that entrained or ablated material can exhibit different redshift progressions.)

D. QUANTIZED REDSHIFTS IN THE REST FRAME OF THE PARENT GALAXY

The Karlsson redshift quantization values were found for bright quasars mainly associated with low-redshift galaxies. But when associations are found with fainter apparent magnitude quasars around higher redshift galaxies, one has to correct for the redshift of the parent galaxy. To find the redshift of the quasar in the reference frame of the ejecting galaxy one needs to divide by the redshift of the galaxy, i.e.,

$$(1 + z_Q) = (1 + z_1)/(1 + z_G).$$

Failure to realize this led some to reject the quantized redshift values because "the values drifted away from the peaks for larger samples at fainter apparent magnitudes." But, as cautioned in the early papers, proper analyses in any sample should calculate the redshift values as a function of the galaxies from which they originated.

For ejecta originating from low-redshift galaxies, this correction is not significant. But in the following *Catalogue* there are a number of examples where correction for an appreciable redshift of the parent galaxy moves the corrected quasar redshift onto, or much closer to, the expected peak redshift. I often point out these cases because I argue that if the quasar were not associated with the galaxy the correction would not, in general, move it onto the peak. If the correction moves the redshift particularly close, this is evidence for the physical association of this particular galaxy and quasar.

E. Hierarchical Associations

Finally there is the question of how large and nearby associations can get. Can one always trace back an even earlier origin from an even brighter, lower redshift galaxy? M 101 is a case in point, with higher red-shift galaxies and secondary and tertiary ejections extending out possibly to a radius of ~15 deg. The brightest objects, like M 101, of course, are generally the closest to us and subtend the greatest angles on the sky. Are fainter associations less luminous hierarchical generations within the nearby structure or are they more distant copies of the nearby system? The evidence that many objects previously believed to be at great distances are actually much closer confronts us with the most drastic possible revision of current concepts. Because this point needs to be explored in more detail, M 101 is treated in a separate section at the end of the *Catalogue* (Appendix A).

One specific question to which this leads is: How many of the observed extragalactic objects in fact belong to the Local Supercluster and how many lie beyond? In this regard, readers will notice that cases of association in this *Catalogue* tend to avoid that part of the sky in the general direction of the center of the Local Supercluster. Clearly that is where the highest density of potential parent galaxies in the sky is encountered. It is suggested, however, that the associations are too intermingled there, and that they simply stand out more clearly in areas of sparser bright galaxy population. This is one of the more difficult questions that is left for future analysis.

References

Arp, H. 1998, *Seeing Red: Redshifts, Cosmology and Academic Science* (Montreal, Apeiron)

Arp, H. 1999, *ApJ* 525, 594

Arp, H. 2001, *ApJ* 549, 802

Burbidge, E.M., Burbidge G.R., 1967, *ApJ* 148, L107.

Burbidge, G.R. 2001, *AJ* 121, 21.

Catalogue Entries

There are three bright X-ray sources within 60 arcmin aligned across NGC 7817, as shown in the adjoining figure (filled circles marked with an x). The minor axis of N7817 is p.a. = 135 deg., just in the direction of Mrk 335. One of the absorption line systems in Mrk 335 is at $z = .0076$, compared to $z = .0077$ for N7817. A compact blue X-ray object lies about half way along the line to the $z = 1.11$ quasar (at $00^h02^m53.5^s$ $+21^d01^m10^s$). A fourth galaxy is near this same line, just out of the frame to the NW. It is an IRAS, UV excess galaxy of $z = .035$ and forms a good pair with Mrk 335 at $z = .026$.

A cluster of X-ray sources lies about 40 arcmin NE of Mrk 335 and to the ESE of NGC 7817. It contains a rather dense group of quasars in the region indicated in the figure by dashed lines. They include the rather intriguing redshifts of $z = 1.38$ and 1.40 and $z = .77$ and $.75$. There is also an X-ray luminous quasar at $z = .389$ in this group which is an almost perfect match for the nearby ASCA quasar of $z = .388$ which is shown in the figure.

Mrk 335 is surrounded by many UV bright objects. It arouses curiosity as to what the fainter ones might be.

ALIGNED OBJECTS

$z = .026$	Sey1	$V = 13.85$	Mrk 335, 1RX
$z = .388$	QSO	$V = 18.1$	EXOSAT 0003.4+2014
$z = 1.106$	RSO	$O = 19.1$	TEX 2358+209, 1RX
$z = .035$	Gal	$m = 15.3$	IRAS F23569+2108

NEEDED

Deep wide field images, identification of UV objects, some redshifts.

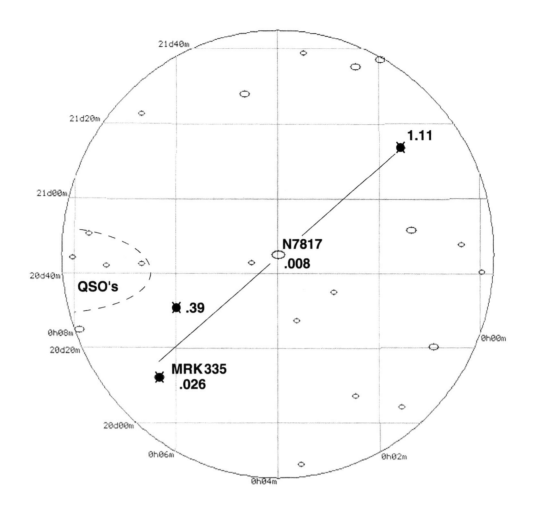

21d40m
21d20m
21d00m
20d40m
20d20m
20d00m

1.11

N7817
.008

QSO's

.39

MRK335
.026

0h08m
0h06m
0h04m
0h02m
0h00m

The Perseus-Pisces filament is very large, stretching across almost a quadrant of the sky. The figure here shows a portion of the western part. Sub-filaments are conspicuous in the figure, but the interesting aspect is that along the major filaments of z = .023 galaxies there are, defining the same filaments, galaxies of z = .08 to .10. There are some groups and clusters along these lines: for example Abell 21 at z = .095, which appears to stretch up to the Zwicky cluster at z = .023 in the center of the field. There are also an X-ray BSO and a z = .51 Seyfert, which may be associated with the latter filament. Near the NGC 68 group there is a BL Lac object with no present redshift.

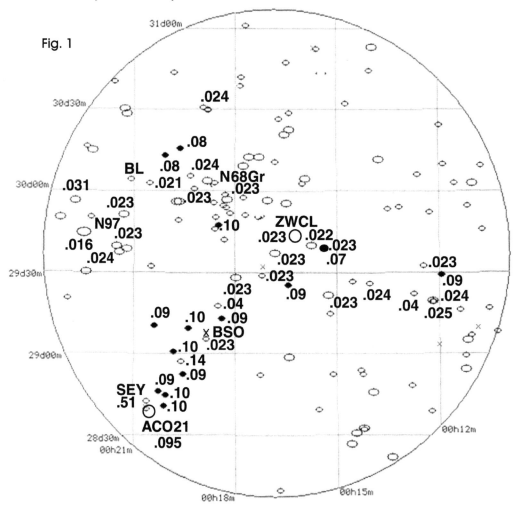

Fig. 1

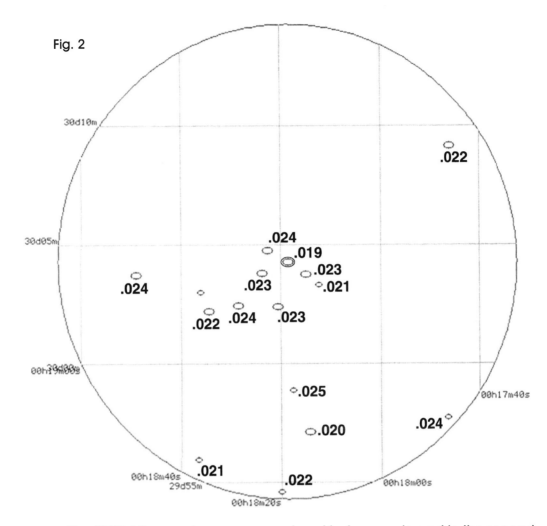

Fig. 2

The NGC 68 group is very compact and is shown enlarged in the second figure here. It is seen to be near the center of a line of four companion galaxies, all of which are somewhat higher redshift. The rest of the group is of higher redshift, presenting another example where the parent or dominant galaxy in a physical group is some hundreds of km/sec lower redshift. In this case the group would average about 1200 km/sec higher.

Galaxy filaments are accepted as common features in the sky. Filaments of much different redshift which are superimposed in detail, as the present ones, however, contradict the basic assumptions about redshift distances and should be subjected to all possible observational tests.

NEEDED

Deep wide field images, redshifts of remaining galaxies and objects in the filaments.

NGC 214

$00^h\ 41^m\ 27.9^s\quad 25^d\ 30^m\ 01^s$

$B_T = 13.0$ mag. $z = .015$ Scl: Perseus-Pisces Filament

The X-ray cluster CLG J0030+2618 contains an X-ray luminous galaxy of $z = .516$. The CRSS[*] measures of X-ray sources in the PSPC[†] observations which include this cluster reveal a number of quasars in the range mostly from $z = .5$ to 1.7. The question arose as to where the parent galaxy to this unusual grouping might be. As the accompanying figure shows, there is a line of galaxies leading from NGC 214 directly toward the cluster.

The remarkable feature of this line of galaxies is that their redshifts increase monotonically as they leave NGC 214 and approach the cluster ($z = .015, .02, .. .03, .. .07$). The cluster itself has a wide range of quasar redshifts, and to quote Brandt et al., *AJ* 119, 2359, an X-ray *background source density more than 4.5 times expected*! This strongly suggests an aggregate of different redshifts at the same distance.

Looking in the opposite direction from NGC 214, there is, amazingly, a string of galaxies leading to another cluster of galaxies, Abell 104. The same kind of progression of redshifts is observed ($z = .015, 03,.. .05, .. .04, .. .05, .. .08, .. .09$). The Abell Cluster has $z = .082$. The Abell cluster itself is also extended along the line back to NGC 214 as if its original material had been ablating as it moved outward. For further discussion see Appendix B, Ejection Figs. 9, 10 and 11.

NEEDED

Deep wide field images, redshifts of X-ray sources and further galaxies along the lines.

---

[*] Cambridge ROSAT Serendipity Survey (*Mon. Not. Roy. Astr. Soc.* 272, 462, 1995).
[†] ROSAT, low resolution X-ray photon counter.

52

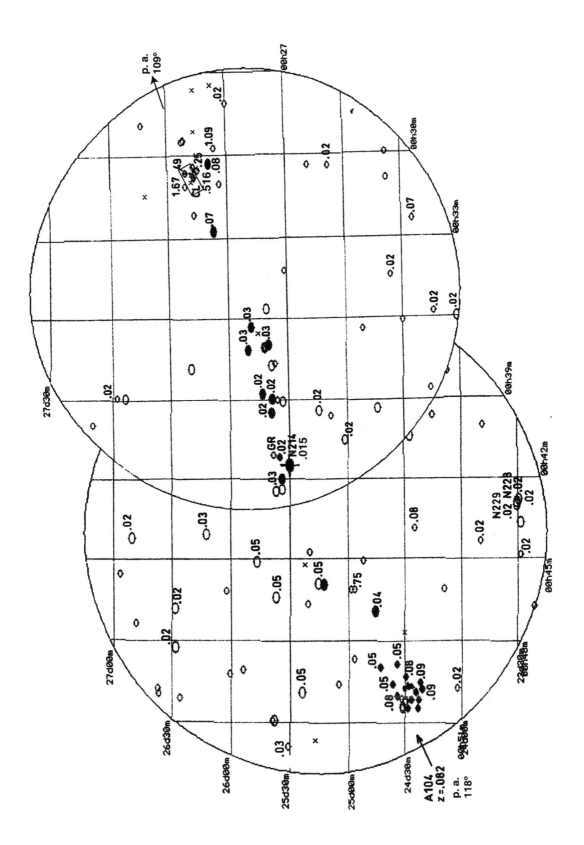

ESO 476-20

01^h 31^m 40.0^s −25^d 32^m 41^s

V = 15.0 z = .020 E gal, IRAS Source

The bright apparent magnitude quasars at 16.6 and 17.6 mag. are aligned accurately across the central galaxy which is also an infrared source. They are also equally spaced at 1.14 and 1.18 degrees. The SE quasar is a strong radio source and the NW quasar is probably to be identified with the strong, 1RX, X-ray source.

Closely along this line are a number of objects indicated in the adjoining plot. Firstly there are four galaxies all of z = .071 and .070. This is close to the quantized value of z = .062. Also aligned are Abell Clusters 214 and possibly 210. Abell 214 is apparently a strong X-ray cluster and there is an all sky survey X-ray source falling along this line which is identifed as a blue stellar object of R = 17.4 mag. and is a good quasar candidate.

In the accompanying figure there is a brighter galaxy plotted to the east of ESO 476-20 which has a redshift 229 km/sec less. This suggests that it is the parent galaxy of ESO 476-20 and that the latter is its active companion. Along this E-W line are found two clusters of fainter galaxies and a gamma ray burster (GRB). On the other side is Abell 206, all of which may indicate a second line of ejection of somewhat different kinds of objects (see ejection lines from M 101 in Appendix A).

ALIGNED OBJECTS

z = 1.20	QSO	V = 16.6	GD1357,1RX:
z = 1.53	QSR	O = 17.6	1Jy 0133-266
z = .160	GCl	m_{10} = 17.5	ACO 214,1RX
z = .---	BSO	R = 17.4	1RXS

NEEDED

Spectrum of BSO, direct images of clusters, spectra of additional objects along alignment.

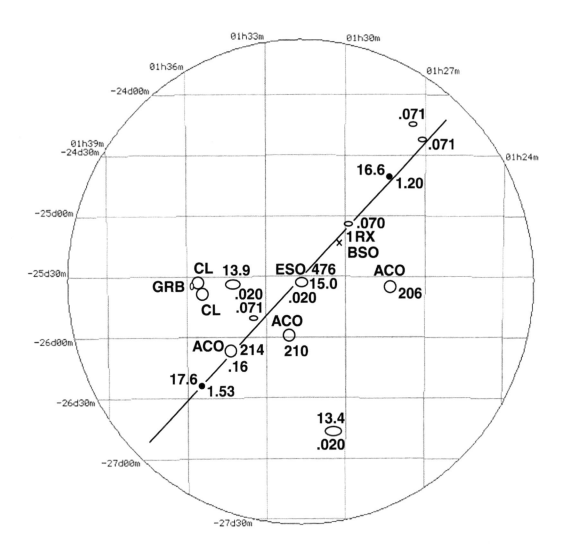

This is a new and particularly striking example of ejected quasars from an active galaxy and supports the long standing suggestion that spiral arms are caused by compact ejections from the nuclei of sprial galaxies (Arp, *Sci. American*, 1963).

The radio map (Fig. 3, NVSS, Condon et al. 1998) indicates ejection along the narrow bar of this active galaxy. The bar and this presumed radio ejection (possibly two sources on either side) point directly to quasars from the 2dF survey of z = 2.22 and z = 2.06 as shown on the accompanying map. As for the quasars further out, they also tend to be placed along the ejection direction, particularly the z = 1.86 and z = 1.48 quasars. So there are two sets of quasars suggesting a match with the two inner pairs of radio sources. The numerical value of the quasar redshifts also argue against their being accidental background projections. We note particularly the the average redshift of the nearest two on the NW as being z = 2.04 closely matching the z = 2.06 of the ones on the SE. The two in the outer most pair are also very similar at z = 1.41 and z = 1.48.

There is a quasar 13.6' E of NGC 613 in the Fig. which is very bright at 15.7 apparent magnitude and is a very strong X-ray source. Quite closely aligned on the other side of NGC 613 is a gamma ray burster, 4B 940216, (located 17.2' W, just beyond the z = 1.69 quasar in the Fig.). This pair of unusual objects fits the distance and alignment criteria of ejected objects from bright galaxies. Association of GRB's with active, low redshift galaxies has been pointed out in this *Catalogue* and by G. Burbidge (2003).

It is also notable that the bright quasar has z = .699 and only 17" N of it there is a quasar of z = 1.177. When the latter redshift is transformed into the rest frame of the brighter object it becomes z_0 = .28—very close to the Karlsson preferred redshift peak of z = .30. If physically associated it would suggest a secondary ejection.

Fig. 2 shows a IIIa-J Sky Survey picture NGC 613, a multi arm spiral. Images of the interior regions and infra red wavelengths show a narrow bar running in the direction of the two nearest quasars.

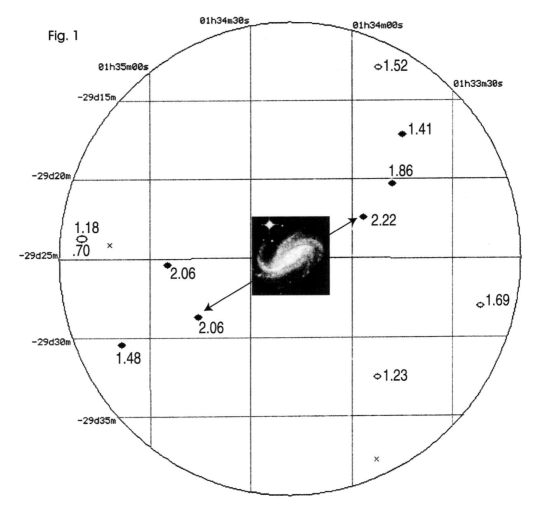

Fig. 1

NEEDED

Higher resolution, deeper, radio and optical mapping of the narrow bar in NGC 613.

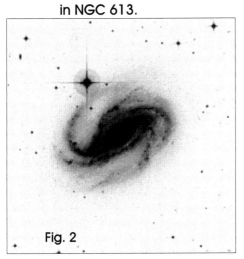

Fig. 2

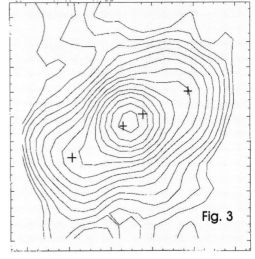

Fig. 3

NGC 622

m$_{pg}$ = 14.1 mag. z = .017 SBb, Mrk 571, em line
with straight arm to QSO

and UM 341

01^h 34^m 18.2 00^d 15^m 37^s

V = 16.9 mag. z = .399 Sey1, QSO, PHL 1037

The great importance of this particular field is to show how large scale surveys of quasars should be examined in order to study the redshift peaks at the Karlsson values of ... .30, .60, .96, 1.42, 1.96, .2.64, 3.48 ... We start by looking at NGC 622, an active Markarian galaxy with two quasars only 71 and 73 arcsec away. At such close separation it is clear that they are associated, even if one did not notice the straight arm coming out of the galaxy and ending almost on the z = 1.46 quasar. (see *Quasars, Redshifts and Controversies*, 1987, Interstellar Media, Berkeley, p. 9-11.)

But this is part of a region which has recently been scanned for quasars in the *Sloan Digital Sky Survey* (SDSS). The quasars now known are shown in the accompanying figure with their redshifts written next to their plotted positions. One first notices that there is a loose group of quasars about 35 arcmin SW of NGC 622 (filled circles). Four of them have closely the same redshift, between z = .72 and .78. Generally, inside this group are higher redshift quasars, all roughly centred on a relatively bright, active Sey1/QSO. But, distressingly, all nine of these quasars fall not only off the predicted peaks—but fairly exactly *between* the peaks. If one transforms them, however, to the reference frame of the central UM 341, which has z = .399 they magically fall remarkably close to the expected peaks!

QUASARS ASSOCIATED WITH UM 341

Name	mag. (g)	z	z$_0$	Δz peak	Remarks
UM 341	16.6	.399			Seyfert parent
SDSS	18.4	1.666	.91	−.05	
SDSS	18.6	.718	.23	−.07	
4C	V = 21.7	.879	.34	+.04	PKS B–V = .84
SDSS	19.0	.745	.24	−.06	
UM	18.2	1.31	.65	+.05	
SDSS	19.3	1.805	1.01	+.05	Radio Gal
SDSS	21.9	3.183	1.99	+.03	
SDSS	19.1	.734	.25	+.05	
SDSS	19.1	.781	.27	+.03	

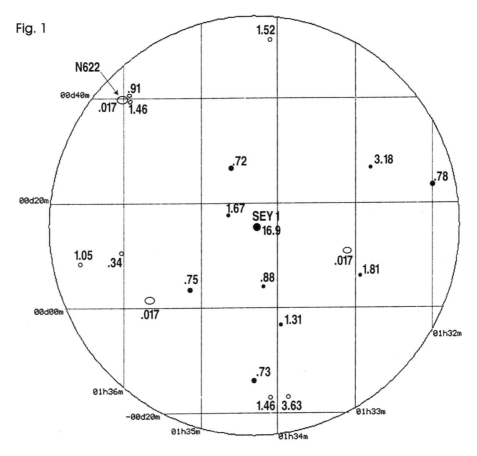

Fig. 1

What about the remaining 7 quasars around the eastern circumference of the field (open circles)? They turn out to fit the Karlsson peaks very well when transformed to the z = .017 galaxies like NGC 622, which are present over the field. It is noticeable that many active galaxies—not only in the Perseus-Pisces filament, but all over the sky—have this ~ 5000 km/sec redshift.

QUASARS ASSOCIATED WITH NGC 622

Name	mag. (g)	z	z_0	Δz peak	Remarks
NGC 622	m = 14.1	.017			Mrk 571
UB1	18.4	.91	.88	−.08	
BSO1	19.0	1.472	1.43	+.02	
SDSS	18.9	1.501	1.46	+.05	
SDSS	19.3	2.749	2.69	+.05	
FIRST	17.8	.344	.32	+.02	Radio Gal
SDSS	18.6	1.522	1.48	+.07	
SDSS	19.2	1.049	1.01	+.05	

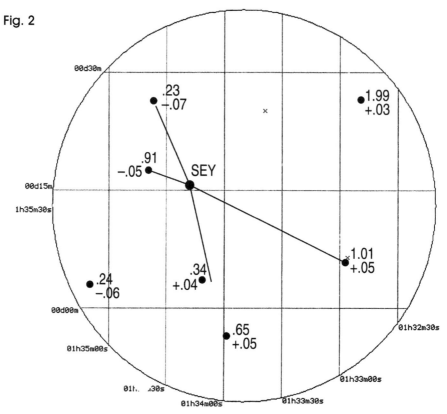

Fig. 2

Labeled are redshifts in rest frame of UM 341 and
delta z's.

*The important point demonstrated above is that if the quasars
are simply analyzed without inspection of the field, no redshift
periodicity will be found. If parent galaxies are identified, the
periodicity will be conspicuously observed. Unfortunately failure to actually look at the data in detail resulted in a press release by Hawkins, Maddox and Merrifield (M.N.R.A.S. 336, L13,
2002) which publicized widely the conclusion that the largest
body of data on quasars proved that there was no quantization present in the redshifts.*

One might ask if UM 341 and NGC 622 are related. It can be seen that
UM 341 falls only about 35 arcmin from the very active NGC 622. This is
typically the distance for ejected QSO/AGN's, and UM 341 then would
represent a case where the QSO was in turn ejecting QSO's. The clinching
evidence in this case lies in the nvss radio map (NGC 622 Fig. 3). There it
is seen that a radio jet stretches about 5 arcmin to the SW at a position
angle about p.a. = 236 deg. Since UM 341 is located at p.a. = 228 deg.
This is strong supporting evidence that the it has been ejected in this di-

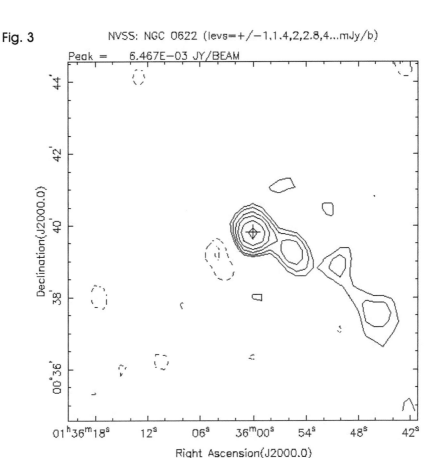

Fig. 3

NVSS: NGC 0622 (levs=+/−1,1.4,2,2.8,4...mJy/b)

Peak = 6.467E−03 JY/BEAM

Radio map of NGC 622 and z = 1.46 quasar.

rection from NGC 622. (The first concentration to the SW in the jet represents the z = 1.46 quasar.) The fact that the QSO's ejected from the younger object (UM 341) are on average only slightly fainter in apparent magnitude than the quasars associated with NGC 622 then furnishes important clues for the nature of the nucleus ejecting the object.

It is very important to note that when the quasars form apparent pairs across the ejecting galaxy, as they do around UM 341, the small deviations from the redshift peaks (listed as Δz peak in Table 1) are usually plus and minus, representing a small velocity of ejection away from and toward the observer: $1 + z_v = (1 + z_0)/(1 + z_{peak})$. This pattern is shown in Fig. 2 here. We will see more examples of this throughout the Catalogue, particularly in the SDSS field around NGC 3023 and the mixed SDSS and 2dF field around UM 602. An example of redshift periodicity at the 10^{-4} level of being accidental will be shown for the 2dF field around NGC 7507.

NGC 632

$01^h 37^m 17.7^s$ $05^d 52^m 38^s$

$m_{pg} = 13.5$ mag. $z = .011$ S0pec, Mrk 1002, IRAS

This central galaxy was brought to my attention by Fernando Patat, who is studying it in detail with the VLT. It is the center of an exactly aligned, equally spaced pair. The SW member of the pair is NGC 631 at $\Delta v = +2,362$ km/sec. The NE member of the pair is a moderately bright quasar at $z = .615$. Since the central galaxy is low redshift, it is expected that the quasar redshift would be near the quantized peak of $z = .60$. But if it is corrected into the rest frame of the $z = .011$, galaxy it comes extremely close at $z = .597$.

There are numerous other PHL candidate quasars in the more extended area around NGC 632.

ALIGNED OBJECTS

$z = .011$	Gal	$m = 13.5$	NGC 632
$z = .019$	Gal	$m = 15.0$	NGC 631
$z = .615$	QSO	$V = 18.2$	PHL1072

NEEDED

X-ray observations of field.

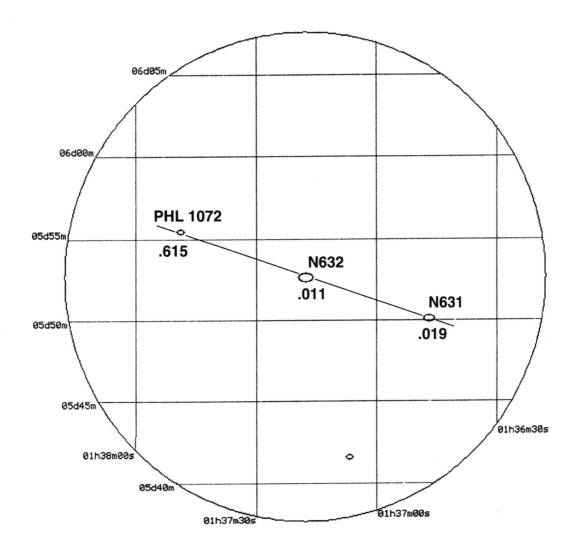

About 14 arcmin SW of NGC 720 is one of the most luminous X-ray clusters known, RXJ 0152.7-1357. The cluster is very elongated, both optically and in X-rays, and points in the direction of the very bright Shapley-Ames Galaxy. (see Arp 2001, *ApJ* 549, p. 816). NGC 720 is active, with an X-ray filament extending from the nucleus and curving southward as it emerges (Buote and Canizares 1996).

There are two further intriguing aspects of this region: One is the two extended X-ray sources, one of which is diametrically opposite to the z = .83 X-ray cluster. Such objects usually turn out to be identified with clusters of galaxies. Are these two clusters? And if so are they elongated? Secondly the Palomar Schmidt deep (dss2) survey seems to show at the limit, two very elongated groups of galaxies, closer to NGC 720 and aligned closely along the minor axis (see dashed contours in the accompanying figure). Possibly associated with the SW candidate cluster is an X-ray source, emission line galaxy of z = .17. A deep, fairly wide field exposure is needed to confirm the important possibility of these two candidate clusters and their possible elongation.

There is a strong radio, strong X-ray source which has been identified with a z = 1.35 quasar and is closely along the extension of the line from NGC 720 beyond the powerful X-ray cluster. It is faint in optical wavelengths (R = 20.4, B = 20.2 mag.). This is typical of Bl Lac type objects and reminds one of the object just N of NGC 4151 and other similar cases (Arp 1997).

From the PSPC observation of this field, a strong X-ray BSO exactly across NGC 720 from the z = .83 X-ray cluster is plotted in the accompanying figure. It is No. 11 in the ROSAT 2RXP Source Catalogue and is identified with a very blue BSO in the next to last line of the table below. (See Appendix B for a position of this X-ray BSO and Ejection Figs. 7 and 8).

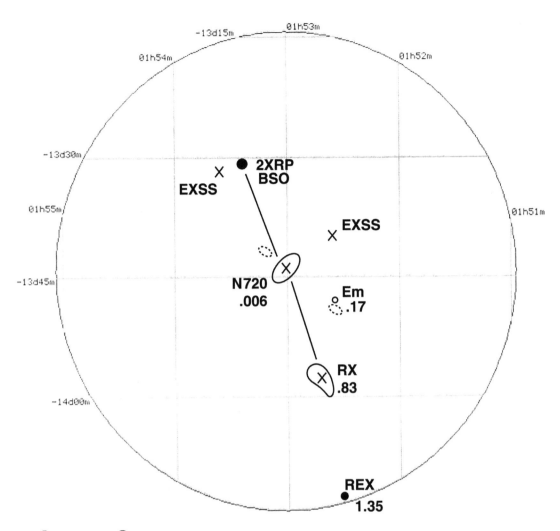

ALIGNED OBJECTS

z = .83	GCl	V = --	RXJ 0152.7-1357
z = 1.35	QSR	R = 20.4	1REXJ015232-1412.6
z = .---	EXSS	m = ?	extended X-ray source
z = .---	EXSS	m = ?	extended X-ray source
z = .---	PSPC	R = 19.0	C = 5.8 cts/ks, BSO
z = .17	emGal	V = 19.2	1RXS

URGENTLY NEEDED

Deep, wide field (about 30 arcmin radius) imaging to identify suspected galaxy clusters in the field, particularly along minor axis of NGC 720. Also redshifts of the X-ray BSO and 4-5 "blue" objects in the field.

The extremely rapid (≤ 1hr) radio variable, QSO 0405-385, is one of a pair of strongly radio emitting QSO's centered on an *unprecedentedly energetic* X-ray emitting (~4000 cts/ks) active galaxy. The most unusual aspect of this galaxy is that even with its bright nucleus, it is optically several magnitudes fainter than active galaxies which have comparably huge X-ray fluxes such as NGC 1275, NGC 4253 and NGC 7213.

The chance of accidental pairing of background objects, despite the wide separation, is ≤ 10^{-5}. In addition, the objects are extremely unusual, very strong radio quasars and X-ray sources, one of them a rapid radio variable with implied excessive brightness temperature and also detected in gamma rays!

There are a moderate number of galaxies in the field, but two have almost exactly the same redshift at z = .059 and are aligned either side of the central Seyfert and closely along the line of the two quasars. They are only slightly greater redshift than A 27.01, at z = .056. They may represent entrained material.

Note: The association was discovered when G. Burbidge noted the report of an unusually fast variation for the quasar and asked Arp if he could suggest a distance for the quasar. The pairing of the quasar with another bright quasar across the unusual Seyfert was then discovered. The configuration was conservatively estimated to have only about a 10^{-5} chance of being accidental. But this did not take account of the extremely unusual nature of the objects involved.

A detailed account of the discovery of this extraordinary X-ray galaxy was submitted to the *Pub. Astron. Soc. Pacific (PASP)*. A referee recommended rejection of the paper on the grounds that it had not calculated the probablility of the configuration being accidental. Arp replied by pointing out the page and paragraph where the calculation had been made. The editors then rejected the paper on the grounds that no changes had been made to accommodate the referee. Arp resigned from the ASP on grounds of editorial misconduct and suppression of scientific information to its members/readers.

I mention this history for two reasons:

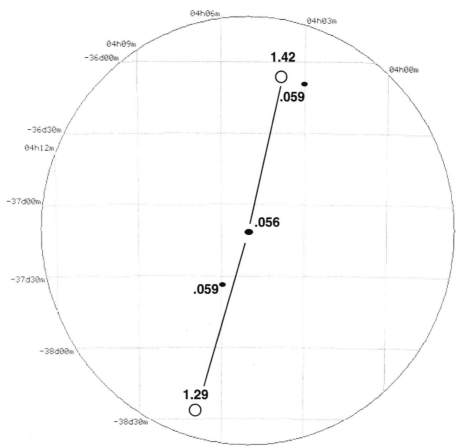

1) To demonstrate that many professional astronomers not only disbelieve the significance of the associations presented in this *Catalogue*, but they oppose their publication.

2) To stress the fact that readers must draw their own conclusions from the data. If the conclusion is that the data is of significance, then the most important question then becomes: "Does academic science require fundamental reform?"

ALIGNED OBJECTS

$z = .056$	Sey1	$V = 15.4$	ESO 359-G019
$z = 1.417$	QSR	$V = 17.2$	PKS 0402-362
$z = 1.285$	QSR	$V = 18$	PKS 0405-385
$z = .059$	gal	$B = 15.56$	APMBGC
$z = .059$	gal	$m = ---$	APMBGC

NEEDED

Deep, high resolution images of the extreme X-ray Seyfert, data on surrounding cluster.

NGC 2435

07^h 44^m 13.7^s +31^d 39^m 02^s

m = 13.5   IRAS source   z = .0140

companion

07^h 43^m 33.0^s +31^d 32^m 06^s

m = 15.7   (UGC 3986)   sf of pair   z = .0125

Two very bright quasars, from among about 100 individually imaged by HST, form an obvious pair across NGC 2435. They are better aligned, however, across a close pair of companions, as shown in the opposite Figure. Companion galaxies are previously identified sources of quasar ejection (*ApJ* 271, 479, 1983). But usually the companions have slightly higher redshift than the main galaxy. In this case the companion whose redshift has been measured has about 450 km/sec less redshift, suggesting it was the originally larger galaxy but fragmented in the process of ejecting the quasars. Supporting this scenario are two higher redshift quasars, still very bright for their redshift, which are aligned exactly across the z = .0125 companion.

The z = .46 quasar is interesting because there are much fainter galaxies around it of roughly the same redshift. But the brightest of the galaxies within this cluster are about .06 less redshifted than z = .46. The fainter galaxies in the cluster, including the quasar, thus appear to have higher intrinsic redshifts. One of these fainter galaxies, however, has z = .607— very close to the redshift of the z = .63 quasar on the opposite side of the NGC 2435 group.

Note: The redshifts of the two brightest quasars across the low redshift galaxy are intriguingly close to the redshifts paired across the famous Seyferts Mrk 205 and NGC 4258:

NGC 4258	z_1 = .40	z_2 = .65
Mrk 205	z_1 = .46	z_2 = .64
NGC 2435	z_1 = .46	z_2 = .63

ALIGNED QSO's

z = .462	RSO	V = 15.63
z = .630	RSO	V = 16.14
z = 1.531	QSO	V = 18.1
z = 1.909	QSO	V = 17.56

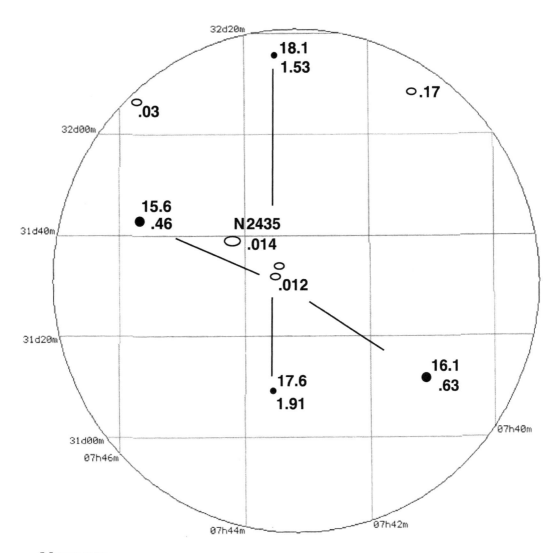

NEEDED

There is only a relatively short HRI X-ray exposure available on the z = .63 quasar, but it records 4 or 5 sources leading accurately back in the direction of the NGC 2435 companion. Deeper X-ray exposures should be obtained on all the quasars.

Mrk 91

08^h 32^m 28.2^s $+52^d$ 36^m 22^s

$m_{pg} = 14.7$ mag. $z = .017$ IRAS source

in Zwicky Cluster 0829+5245 $08^h33^m18^s$ $+52^d35^m$

$m = 13.9$ and fainter, $z = .016$

Two Markarian objects and one UGC galaxy form the core of this Zwicky cluster. A bright apparent magnitude quasar of $z = 3.91$ falls in the cluster. Even when gravitational lensing is hypothesized the quasar is still derived to be among the most luminous known. The *Sloan Digital Sky Survey* (SDSS) screening for quasars with $z > 3.94$ turns up two more close by the $z = 3.91$ quasar at $z = 3.97$ and 4.44.

Mrk 91 is a fairly bright, active galaxy near the center of this Zwicky Cluster, and there are indications of alignments of $z = 2.06$ and .34 quasars to the SW of it plus a tight group of NGC galaxies at $z = .045$ to the NE.

The quasars (filled circles), and Zwicky cluster and NGC 2600 cluster (open circles), at $z = .016$ and .04 are first shown alone in order to emphasize their association. The next Figure adds many of the details of the objects in the region.

ALIGNED OBJECTS

$z = 3.91$	QSO	$R = 15.2$	QSO B0827.9+5255
$z = 3.97$	QSO		SDSS J083324.57+523955.0
$z = 4.44$	QSO		SDSS J083103.00+523533.6
$z = 4.02$	QSO		SDSS J083212.37+530327.4
$z = 2.06$	QSO		CLASS B0827+525
$z = .34$	AGN	$V = 20.3$	87 GB 08241+5228, radio
$z = $ ---	Gal:	$V = 17.5$	1RXS J083010.5+523031

NEEDED

Deep images, spectra of X-ray object, additional galaxies along line.

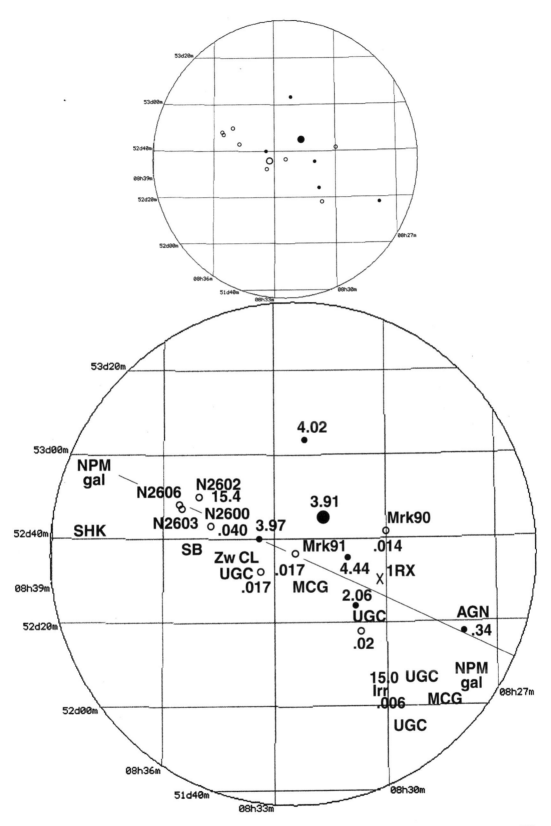

As a sample of what might be contained in the NORAS survey of X-ray galaxy clusters (Böhringer et al. *ApJS* 129,435), it was noted that two RXC clusters of $z = .38$ and .41 fell close together on the sky. The Simbad plot shows that they are contained in a long, conspicuous string of galaxies. Of those that have been measured, there are 7 with $.050 \leq z \leq .057$.

Also along this line, just NE of the $z = .41$ RXC cluster is a radio loud, X-ray Gal/QSO? at $z = .43$. At the NE end of this same line there is an Abell Cluster of $z = .093$ and a neighboring ACO at $z = .095$. Another Abell Cluster, ACO 710 lies nearer the center of the line close to the $z = .38$ RXC cluster.

The only large, active(?) galaxy in this line is NGC 2649 at the end of the line.

ALIGNED OBJECTS

$z = .378$	GCl	-------	RXC J0850.2+3603
$z = .411$	GCL	------	RXC J0856.1+3756
$z = .43$	G:	$O = 19.4$	B3 0854+384, X-ray
$z = .095$	GCl	$m = 16.7$	ACO 727
$z = .093$	GCl	$m = 16.7$	ACO 724
$z = .---$	GCl	$m = 17.9$	ACO 710

NEEDED

Redshifts of more objects along the line, analyses of neighborhoods of the 378 X-ray clusters in the NORAS Catalogue.

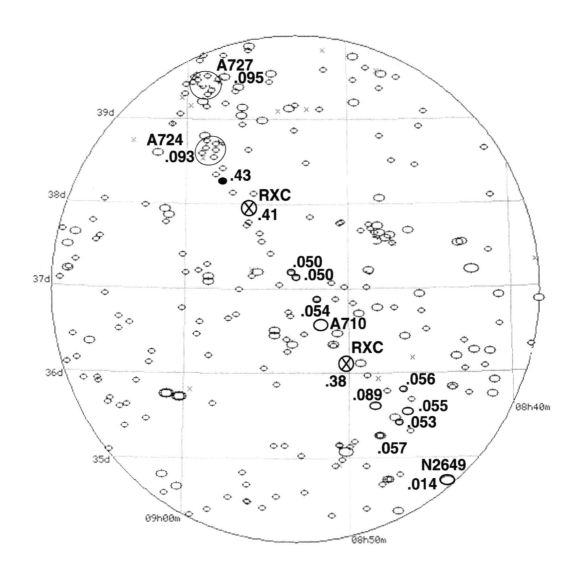

A727
.095

A724
.093

.43

RXC
.41

.050
.050

.054
A710

RXC
.38
.089

.056
.055
.053
.057

N2649
.014

39d

38d

37d

36d

35d

08h40m

09h00m

08h50m

The importance of this particular field is that it again demonstrates that the SDSS and other recent large scale surveys of quasars show many concentrations of quasars which are not only at the redshift peaks of ...30, .60, .96, 1.42, 1.96, 2.64, 3.48... but that these quasars are paired across large and/or active galaxies in such a way as to preclude accidental association.

The Figure shows six quasars in a field of radius 30 arcmin which are centered on the active triplet of galaxies around NGC 3023. The outstanding pair consists of z = .640 and z = .584 quasars which fall z$_v$ = +.02 and − .02 from the major redshift peak at z = .60. It would be difficult to avoid the implication that they had been ejected from one of the central galaxies and were now travelling with a radial component of velocity .02c, one away from, and one toward the observer.

Moreover, there are two other pairs of quasars in approximately the same direction. One pair has apparent velocity deviations from the redshift peaks of +.09 and −.07 and the other +.03 and −.06. This pattern is characteristically encountered (e.g., see UM 341 in the NGC 622 field and pairs analyzed in the introduction). The chances would seem vanishingly small to find repetitions of such patterns in random associations of background objects. In the present case, however, there is even more evidence for association in the fact that both members of the major pair at z = .60 are strong radio sources. The central galaxies are both NVSS radio sources and the *the two quasars are each double radio sources*. The latter is quite unusual and represents additional evidence against accidental association.

It should be noted that the central galaxies here are low redshift so that only small corrections to their rest frames are needed. But in cases where the ejecting galaxy has appreciable redshift it is critical to correct the observed redshifts. Failure to do this has led to some well publicized claims of non-quantization of quasar redshifts. There is now available, however, a computer program by Christopher Fulton which plots on the surveyed sky regions, quasars in specified redshift intervals. The few samples presented in this *Catalogue* indicate that the surveys contain a rich

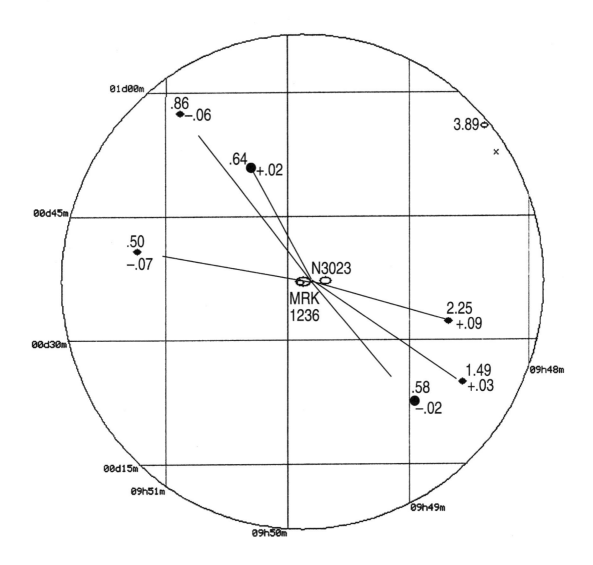

harvest of associations with the potential of yielding data on the associations the nature of the redshifts.

NEEDED

Spectroscopy and imaging of the Mrk object which is involved in the disturbed spiral arm of NGC 3023.

75

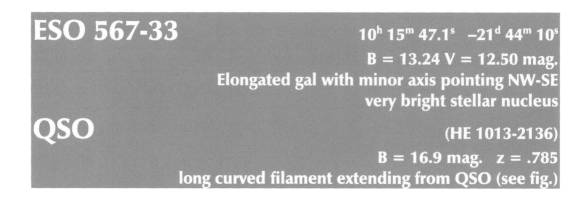

ESO 567-33

$10^h 15^m 47.1^s$ $-21^d 44^m 10^s$

B = 13.24 V = 12.50 mag.
Elongated gal with minor axis pointing NW-SE
very bright stellar nucleus

QSO

(HE 1013-2136)

B = 16.9 mag.   z = .785
long curved filament extending from QSO (see fig.)

This morphologically unusual, bright quasar is roughly along the minor axis direction from a nearby bright galaxy. Much more widely separated, but still along this line are a pair of quasars with closely matching redshifts.

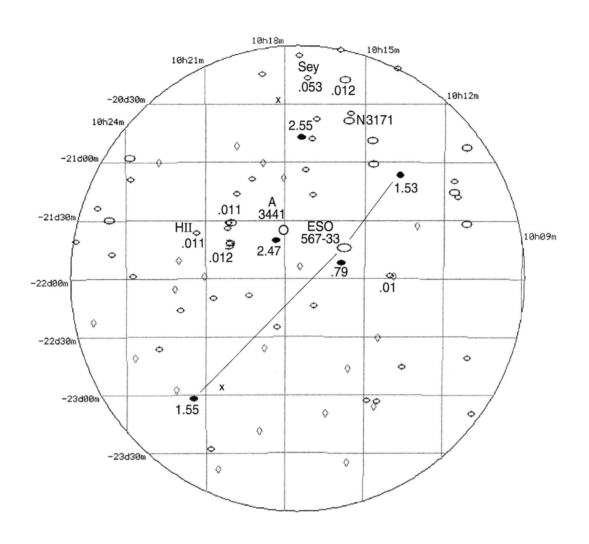

ALIGNED QSO's

| z = 1.53 | QSO | O = 18.1 | CTS J03.17 |
| z = 1.55 | QSO | O = 20.9 | MC1019-227 |

OF INTEREST

z = 2.47	QSO	R = 17.4	CTS J03.14
z = 2.545	QSO	R = 16.7	CTS J03.13
z = ------	GCl	m = 17.4	Abell 3441

NEEDED

1) Spectra and images of ESO galaxy

2) Redshift of Abell Cluster

MCG+01-27-016

$10^h\ 33^m\ 28.1^s$ $+07^d\ 08^m\ 05^s$

m = 15.2   z = .044   IRAS source
(double galaxy in a small group aligned NE)

A supposed gravitational lens of z = .599 has a z = 1.535 image only 1.56 arcsec away. The MCG galaxy is 3.6 arcmin distant and is an IR source. Other objects are aligned on either side of this apparently active galaxy, including radio sources (the lens also). Note that in the reference frame of the z = .60 object the z = 1.535 quasar would represent an emitted object of z = .585, both being close to the z = .6 quantized red-shift peak.

ALIGNED OBJECTS

z = .599	lens	V = --	EQ 1030+074
z = 1.535	compn	V = --	
z = .--	gal	m = --	NPMIG gal
z = .--	rad		87GB rad source
z = .--	rad		87GB rad source
z = .--	EUV	m = --	EUV emission

NEEDED

Deep images and spectra of objects in alignment.

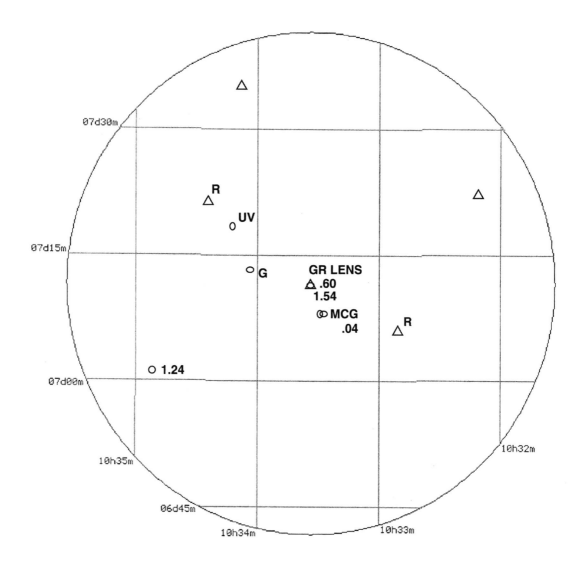

In 1991 a quasar was discovered at V = 13.86 mag. which was "among the three brightest quasars in the sky" (*A & A* 247, L17). Inspection of the region reveals a 14th mag. Seyfert galaxy 52 arcmin away with some indication of alignment of objects back toward the Seyfert. There is a group of X-ray sources around the very bright quasar. There are also some fainter galaxies of the same redshift which form a group or cluster around it. Recent measures of the 1REX source (radio emitting X-ray source) show that it is a BL Lac type object at z = .367. *The z = .367 BL Lac and the z = .393 quasar fall accurately aligned across the bright, z = .086 quasar.* The latter redshifts, in the reference frame of the bright z = .086 quasar, would fall closer to the Karlsson, preferred redshift peaks.

Note: In the following Fig. 2 the original map of this field is shown. At that point it looked as if all the quasars originated from the z = .016 Seyfert, But when the southernmost X-ray source was measured it turned out to have a z = .367 (*ApJ* 556, 181, 2002). This clearly changed the interpretation to a pair of quasars in a secondary ejection across the bright QSO as shown in Fig. 1. The redshifts in this pair then came closer to the expected periodicity as well. The point in showing the development in this particular case is that sometimes just one single measure of an object in some of the fields can strongly reinforce the association and/or clarify the interpretation.

Additional comments can be made that in the reference frame of the z = .016 Seyfert, the (presumably) ejected bright quasar has a z = .069. This is close to the lowest quasar quantization of z = .062. It is similar to the famous case of NGC 4319/Mrk 205 where the latter QSO/AGN has z = .070. Mrk 205 has ejected quasars of z = .46 and .64. We might also note that z = .06 to .07 is a frequent redshift of X-ray galaxy clusters and galaxies in lines such as from M 101 and other cases in this *Catalogue*.

Finally, as to the question of a quasar ejecting other quasars instead of active galaxies ejecting quasars, the definition that separates a "quasar" and an "AGN" at −23 mag. has no operational meaning because it is

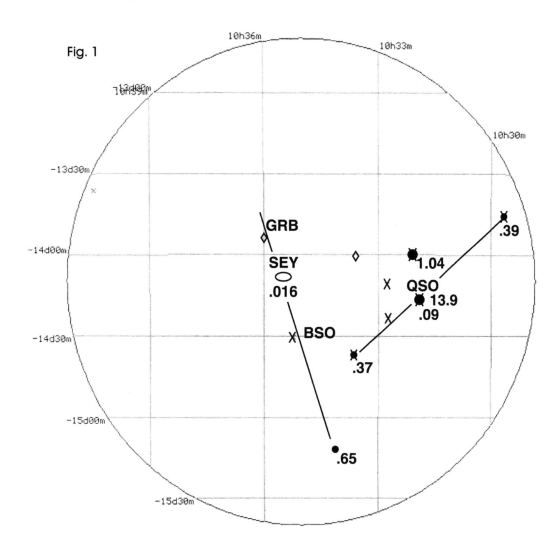

Fig. 1

based on an assumption about redshift distances. Spectral character-isitics are continuous. Moreover an object which varies above and below the arbitrary luminosity cannot change discontinuously from one kind to another.

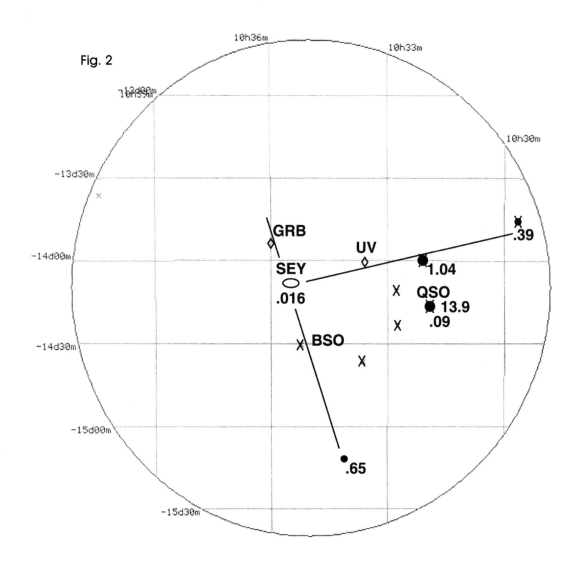

Fig. 2

ALIGNED OBJECTS

z = 1.039	Blazar	R = 18.4	TEX 1029-137
z = .393	QSO	V = 18.3	RXJ102938-134620
z = .367	BLLac	B = 20.2	1REX J103335-1436.4
z = .652	QSO	O = 17.2	HE 1031-1457
z = .--	GRB		GRB 4B950226
z = .--	BSO	R = 19.1	1H 1032-14.2

NEEDED

Spectrum of X-ray BSO, optical identification of X-ray sources, checking of UV sources in the area.

NGC 3312 is the nearest candidate for the dominant galaxy in the famous Hydra Cluster of Galaxies (ACO 1060). The center of the cluster lies about 5 arcmin NW, is very dense and a strong X-ray emitter. The plot shown here has radius of only 10 arcmin. The brighter galaxies listed in Simbad are shown with their redshifts noted where known. Some galaxies of roughly the cluster redshift extend to the SW, reinforcing the impression that NGC 3312 is the center of a cluster whose redshifts range from $z = .009$ to .019.

The fainter cluster members are close enough in redshift to qualify conventionally as physical companions to the central galaxy. But in every case in the Figure they are from slightly to about 3,000 km/sec larger. This is another demonstration of the excess redshifts of companions in groups which has been reported extensively over the years (*ApJ* 430,74, 1994; *ApJ* 496,661, 1998). I cite this here as an example of a rich galaxy cluster showing the same effect. It is a prediction of this *Catalogue* that when more rich clusters are studied in a redshift-apparent magnitude diagram, they will systematically demonstrate this intrinsic redshift effect. (*Note:* There are now many more redshifts known in this cluster. A few are near or slightly less than the redshift of NGC 3312, but most are higher.)

There is a strong pair of radio sources across NGC 3312 and also a strong pair of infrared sources, both roughly within the frame we have shown here. A little more than 10 arcmin SSE of NGC 3312 is a spectacular pair of overlapping spirals (NGC 3314 A and B) with redshifts of 2,872 and 4,426 km/sec. Such differences in redshift for apparently interacting objects are reminiscent of Stephan's Quintet. Perhaps the same explanation obtains here—that they are of different ages but were ejected along the same path from the Sab NGC 3312 (this origin was suggested for the NGC 7331(Sb) in the Quintet).

I also note that within the large extent of the Hydra Cluster (over a degree) there are some subclusters which form a tight line of higher redshift galaxies. (Information courtesy of Daniel Christlein.)

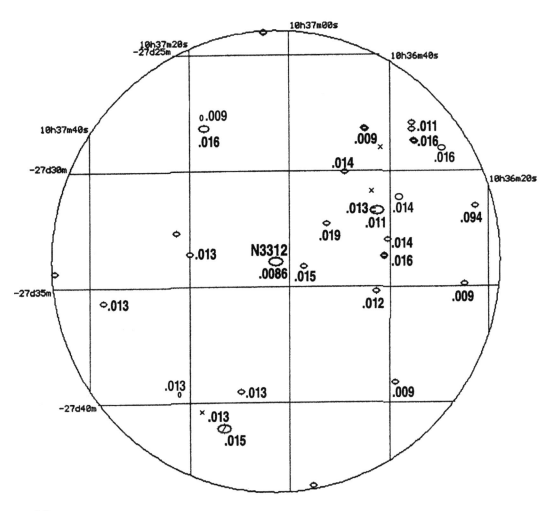

Needed

Studies of more clusters around dominant galaxies.

I am grateful to Alberto Bolognesi for calling my attention to this association.

This large, bright spiral is nearly edge on and has a cluster of galaxies (ACO 1209) extending from its projected edge northward along its minor axis. There is a radio source reported between the cluster and the galaxy. The absorption is so peculiar, cutting sharply through the northern half of the galaxy, that one might consider the possibility of dust from the cluster obscuring part of the galaxy. Intriguingly, there are also two EXSS (extended X-ray sources) to the SW. Such sources often turn out to be faint X-ray galaxy clusters.

Less than 40 arcmin, also due north, are four quasars ranging from $z = 2.14$ to 2.49. They are part of a Weedman, CFHT objective prism field, but represent about an 8 times increase in average density (even though that survey's average density includes some fields centered on active galaxies and some low density fields discarded from the average).

ALIGNED OBJECTS

$z = .--$	Cl	$m_{10} = 17.2$	ACO 1209
$z = .--$	extend X-ray		EXSS 1111.2+1303
$z = .--$	extend X-ray		EXSS 1110.8+1253
$z = 2.14$ to 2.49		QSO's	Weedman objective prism field

NEEDED

Deep images of exterior of NGC 3593, A 1209 and EXSS's. Redshifts of cluster galaxies.

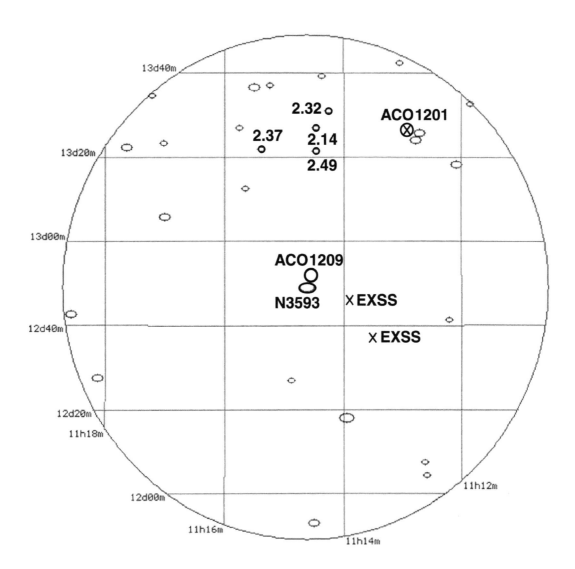

UGC 6312

The most interesting object in this field is a supposed gravitationally lensed object with components ranging from z = 1.718 to 1.728. This group of quasars, however, is located only about 6 arcmin away from a 14.7 mag. UGC galaxy. About 7 arcmin on the other side of the UGC galaxy is a strong X-ray (1WGA) quasar of z = .70. After this initial pattern was noticed, additional quasars were discovered, paired at about 8 and 9 arc minutes each across UGC 6312 and having very similar redshifts of z = .81 and .85.

The supposed lensing galaxy, m = 19.0, z = .31, is in a group of galaxies which forms perhaps the original pair with the z = .70 quasar across UGC 6312.

ALIGNED OBJECTS

z = .698	S1/QSO	V = 21.5	1WGA
z = 1.728	QSO	V = 16.22	PG A(grav lens?)
z = 1.722	QSO	V = 17.6	PG C(grav lens?)
z = 1.722	QSO	V = 18.1	PG B(grav lens?)
z = 1.618	QSO	R = 19.2	1WGA
z = .812	QSO	R = 19.8	1WGA
z = .847	AGN/QSO	R = 20.7	1WGA
z = .--	galaxy	m = --	IR source
z = .31	G	m = 18.96	YDG81

NEEDED

Better spectra of UGC 6312, IR galaxy, deep images and identification of further X-ray sources.

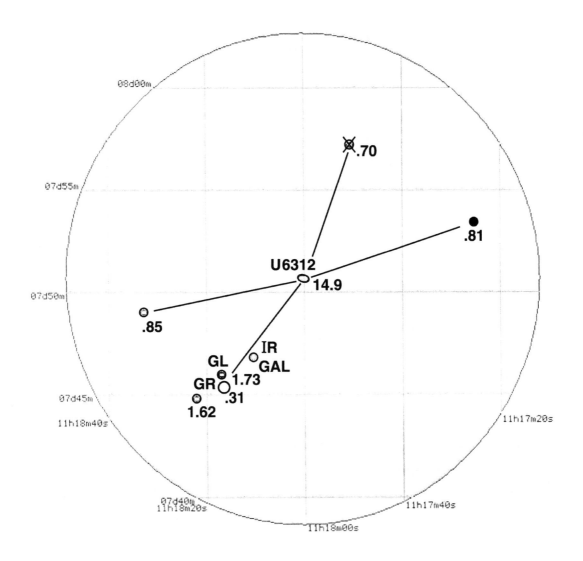

$11^h 44^m 52^s \quad -09^d 14^m 01^s$

$B_0 = 12.3 \quad z = .018 \quad$ disturbed morphology

A very high redshift QSO ($z = 4.15$) lies about 1.5 deg. NNE of the bright, disturbed NGC 3865. Very closely along this same line is a binary quasar from a bright radio survey with $z = 1.342$ and 1.345. Also along this line is the galaxy cluster A 1375. About the same distance on the other side of of NGC 3865 is the cluster A 1344 which has its 10th brightest galaxy at exactly the same $m_{10} = 16.6$ mag. They form an obvious pair across the NGC galaxy. Along the alignment on this same SSW side lie two strong, survey X-ray sources. One can be identified with a bright, blue stellar object, and the other with a medium bright, blue galaxy. A supercluster (SCL, Einasto) lies slightly further along this line. Also, of the galaxies lying along this line, two which have redshifts are plotted with $z = .020$ and $.022$, implying physical association with the central NGC 3865. Bright QSO's (17.1 mag., $z = .425$, PKS, 16.2 mag., $z = .554$, and A 1348, $m_{10} = 17.0$ mag.) lie SSW, out of the frame of the plot and may also be associated.

The strongest support for the physical association of these different redshift objects comes from the PSPC observations of X-ray sources around the NNE quasars which the accompanying plot shows are distributed in an elongated pattern toward the proposed galaxy of origin.

ALIGNED OBJECTS

$z = 4.15$	QSO	$R = 18.6$	BR 1144-0723
$z = 1.342$	bin QSO	$V = 18.7$	PKS 1145-071
$z = 1.345$	bin QSO	$V = 18.7$	PKS 1145-071
$z = .--$	ACO	$m_{10} = 16.6$	A1375
$z = .--$	BSO	$V = 17.8$	1RXS
$z = .--$	Bcg	$V = --$	1RXS
$z = .076$	ACO	$m_{10} = 16.6$	A1344

NEEDED

1) Spectra of 1RXS candidates

2) Redshift of A1375 and deep images A1375 and A1344

3) Deep X-ray and optical images of objects along line

4) Redshifts of more galaxies along line (and in field).

ROSAT
X-RAY
PSPC

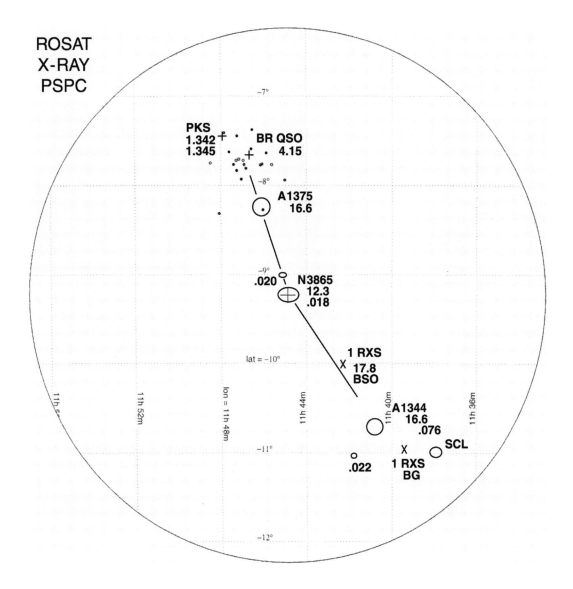

Spectroscopic study of this bright Sey/Quasar has suggested outflow of material. Radio observations first indicated the ejection of two quasars closely spaced across it. (2.6 and 5.5 arcmin distance—See *A & A* 296, L5, 1995 and *Seeing Red* pp15-17.) In the process of later quasar surveys, however, it was never noticed that another pair of bright quasars was *exactly aligned* across this same central object (15.1 and 27.7 arcmin distance). This pair matches so closely in redshift at z = .87 and .85 that it is extremely unlikely to be chance. In fact it is similar to the z = .81 and .85 pair across UGC 6312 (earlier in *Catalogue*.)

The quasars aligned here are all rather bright in apparent magnitude and exceptionally strong in X-rays (C = X-ray intensity in counts per kilosecond.) This would strengthen the association of the central PG Seyfert with the bright M 87 in the Local Supercluster, as outlined in *Seeing Red* (p. 118) and *Quasars, Redshifts and Controversies* (p. 159). In the figure presented here the line of the famous X-ray radio jet which emerges from M 87 at p.a. = 290 degrees and cuts through M 84 on the way is shown here to pass rather closely through PG 1211+143 and along the lines of quasars which appear to be ejected from the active Seyfert.

It is interesting to note that the redshifts become successively lower as the pairs extend further from the central object. This agrees with the pattern observed in NGC 3516 and NGC 5985 (Figs. 6 and 7 of the Introduction) and supports the interpretation that the intrinsic redshifts decrease as the quasars age.

ALIGNED OBJECTS

z = 1.28	QSR	E = 17.0	Radio (4C 14.46), (C = 15)
z = 1.02	QSR	E = 17.0	Radio (NVSS), (C = 20.7)
z = .723	QSO	R = 18.0	X-ray (C = 12.6)
z = .847	QSO	R = 17.2	X-ray (C = 35.8)
z = .870	QSO	R = 18.3	X-ray (C = 18.1)
z = .---	BSO	R = 19.0	X-ray (C = 11.1)
z = .---	NSO	R = 18.2	X-ray (C = 6.6)

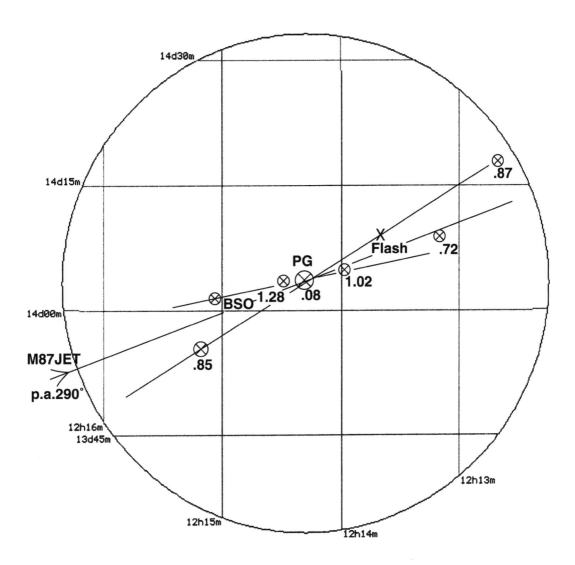

NEEDED

The Blue Stellar Object (BSO) and Neutral Stellar Object (NSO) are in the line of the z = 1.28 and 1.02 quasars and require spectra. Their positions are: $12^h15^m01.82^s$ $+14^d01^m13.0^s$ and $12^h13^m27.7^s$ $+14^d04^m 38.9^s$. The X-ray flash in the figure represents a highly probable event (Gotthelf et al., *ApJ* 466,729, 1996). No optical identification was made, but the position is not highly accurate.

Note: A recent XMM measure (Pounds et al. astro-ph 03036030) shows gas being expelled from the central AGN with velocities of .08 to .10 c. But this just about the ejection velocity (+.08, −.05 c) measured for the pair of quasars at z = 1.28 and 1.02. This would seem direct and quantitative evidence that these quasars originated in the observed ejection.

NGC 4410

$12^h 26^m 28.3^s$ $+09^d 01^m 08.7^s$

m = 13.6 z = .025 Sab interacting with Mrk 1325, Sey 3, as a double nucleus

This very disturbed pair of interacting galaxies—one of which is a Markarian object with compact nucleus—has seven bright quasars within 60 arcmin radius. The closest pair is exactly aligned across the central object and the next closest pair is aligned within 30 deg.

ALIGNED OBJECTS

z = .731	d = 17.5'	m = 18.73	z_0 = .689	mean z_0 = .594
z = .535	d = 23.6	m = 17.87	z_0 = .498	
z = 1.456	d = 25.4'	m = 18.37	z_0 = 1.397	mean z_0 = 1.404
z = 1.471	d = 48.9	m = 17.83	z_0 = 1.411	

When corrected to the central galaxy (z_0), one quasar in each pair should have a component of ejection velocity toward, and the other an equal velocity away, from the observer. *Then the mean of these two redshifts should represent the intrinsic redshift of the ejected quasar. It is seen above that intrinsic redshift is only –.006 away from a Karlsson peak in both cases!* If the test of a physical law is to make numerically accurate predictions then the quantization low of quasar redshifts has just passed an especially crucial test.

Additionally, there is a bright quasar of z = .681 exactly on the same line but further away than the z = 1.456 quasar. It would have a corrected redshift of z_0 = .64.

The NGC 4410 association is a Bonanza in other respects as well. For one, in the closer regions recent Chandra images (Nowak, Smith, Doanahue and Stocke, 2003) have shown that there is a "possible population of ULX sources" (Ultra Luminous X-ray Sources). One is associated with a radio point source and the "brightest ULX may also be associated with radio emission." This would fit perfectly the assignment of ULX's to early stages of ejected quasars where the strong X-ray, radio emission characteristics of BL Lac objects are most prevalent (Burbidge, Burbidge and Arp 2003, A&A 400, L17).

94

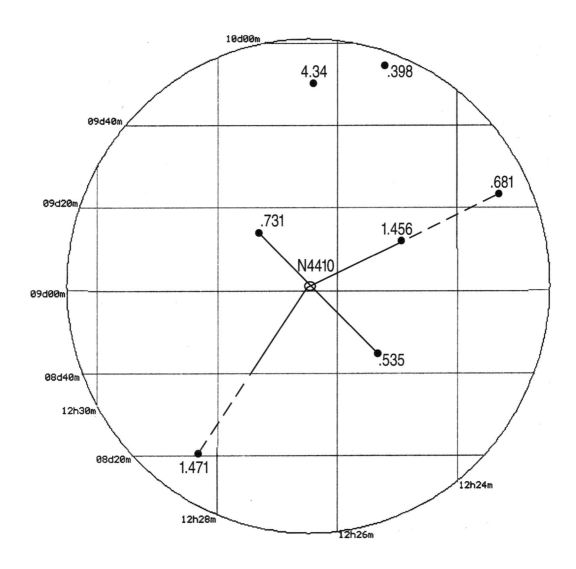

NGC 4410 is projected near the center of the Virgo Cluster but at cz = 7400 km/sec would be conventionally considered a backgound object. However, it appears associated with a large galaxy cluster, Abell 1541 at z = .09, which has been shown to be extended in X-rays away from the putative center of the Virgo Cluster, Messier 49. Along this same line from the active M 49 are a cone of bright quasars which pass over the region of NGC 4410 (*Seeing Red*, Apeiron, 1988, pp. 121 & 142).

There is a line of galaxies and some X-ray sources passing nearly NE-SW through NGC 4410 and the region would be a challenging and rewarding study for more spectroscopic and imaging studies.

This field was located by means of the plotting program of Chris Fulton where a string of quasars with $2.54 \leq z \leq 2.74$ was evident at 13h41m–00d45m. The Sey/QSO was identified at the center of the field and a 40 arcmin radius was searched in NED for quasars with $z \geq 2.5$. Six 2dF quasars were found and they are plotted and labeled in Fig. 1.

It is seen that the closest pair across the active central objects has $\Delta z = +.04$ and $-.10$. The second closest pair has $\Delta z = +.05$ and $-.10$. Overall there are six quasars running NE to SW through UM 602 which fall close to the $z = 2.64$ quantization value (with accidental probabilities ranging from 0.28 to .10.)

Now, however, a very disturbing point arises. If these high redshift quasars are really associated with the central UM 602, their redshifts must be transformed to its $z = .236$ rest frame. That surely must destroy the coincidence with the $z = 2.64$ quantization value. Gritting our teeth we make the transformation, and lo and behold, we calculate $z = 1.92, 1.99, 2.02$ 1.86, 2.01 and 1.86. The mean of these redshifts is $z = 1.94$! The most conspicuous and first discovered redshift peak for bright quasars is 1.96 (originally discovered as 1.95).

One interesting implication is that if the $z = 2.64$ quasars evolve a step down to $z = 1.96$, then UM 602 should step down to near $z = 0$. There are other lower redshift galaxies in the field, but their connection to UM 602 would be a subject for further investigation.

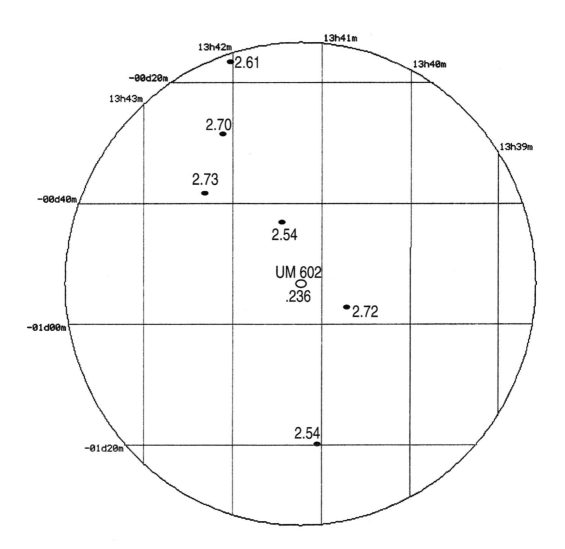

Only X-ray sources are plotted in this radius = 40 arc min plot. As has been noted, Markarian galaxies tend to have numerous X-ray sources nearby which are often elongated in their distribution. In this case there is an elongation of sources to the SW, in the same direction as the optical streamer emerging from this disturbed system. Further along this line is a fairly bright quasar at V = 18.0 mag. and z = .977. Finally at the end of the line is a good optical candiate for an X-ray QSO at R, B = 18.8, 19.0 mag. The X-ray source just to the NE of the latter seems to have no optical I.D.

ALIGNED OBJECTS

z = .977	QSO	V = 18.0	PB 4145
z = 1.196	gal	R = 21.6	RX J135529.59+182413.6
z = .152	QSO	V = 15.71	J135435.6+180518
z = .217	emgal	V = 18.15	1 SAX J13539+1820

NEEDED

Deeper images of central galaxy. Further optical identification of objects SW of Mrk 463 and along the line.

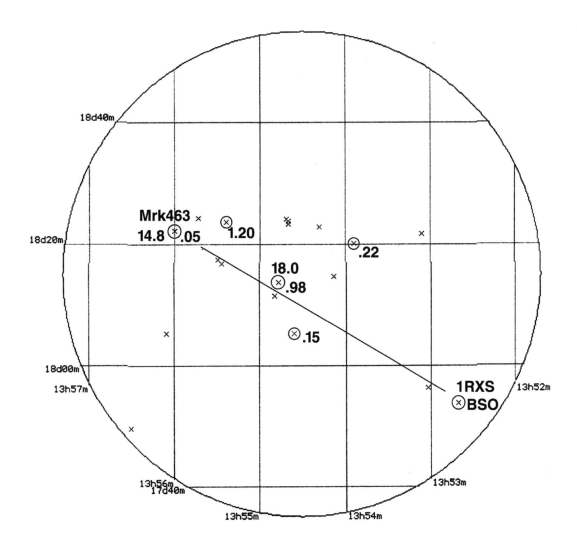

Mrk 478 is very similar to PG 1211+143 (discussed earlier) in having exceptionally strong soft X-ray excess and an abundance of observations and analyses in various wavelengths. Like many other Markarian galaxies it shows lines of X-ray objects emanating from it. The circled crosses in Fig. 2 are predominantly survey (1RXS) sources, and therefore very bright X-ray sources. The blue stellar objects identified with these sources are also relatively bright, and presumably will turn out to be rather bright apparent magnitude quasars.

The innermost pair of candidates has the highest significance, as shown in the High Resolution X-ray map of Fig. 1. Fig. 1 also shows four fainter sources roughly orthogonal to the main pair, a characteristic seen in other examples in this *Catalogue*. The SW identification in the main pair is a blue galaxy, and possibly also a BSO aligned about p.a. = 52 deg. The NE candidate in the pair is accurately aligned, and the two are nearly equally spaced across the central Markarian galaxy.

Further along this direction Fig. 1 shows a remarkable X-ray cluster containing several blue objects that are *elongated accurately back toward the Mrk object* (p.a. = 46 deg.). I consider this strong support for the evidence presented in Appendix B for the ejection origin of galaxy clusters. Just to the E of this cluster are two more strong, bright quasar candidates also aligned back toward Mrk 478. These (and other 1RXS, PCPS and HRI X-ray candidates shown) remain to be investigated. Radio (NVSS) sources seem aligned acrosss Mrk 478.

ALIGNED OBJECTS

z = .---	BGal	V = 15.0	1RXS, ASCA
z = .---	BSO	V = 16.7	close to above
z = .---	BSO	V = 17.6	1RXS
z = .---	Gal	V = ---	Radio(FIRST) gal
z = .---	GalCL	V = ---	1RXS Cluster, BSO's
z = .---	BSO	V = 18.3,18.7	1RXS,Two BSO's
z = .---	CSO	m = 17	Case Blue Object

NEEDED

Spectra of BSO's, deep image of elongated cluster.

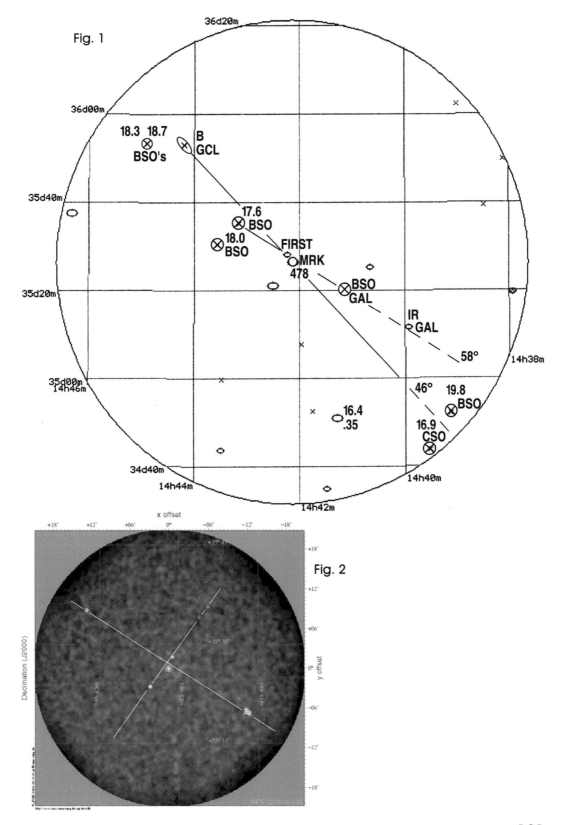

Fig. 1

Fig. 2

101

The supposedly gravitationally lensed quasar SBS 1520+530 is of considerable interest because it is a BAL (broad absorption line) quasar. The configuration is remarkable because there are four images closely in a line. The two brightest have $z = 1.855$ and $z = 1.844$. The two fainter images are reported as stars, although it is not clear that spectra have been taken. As the map within 50 arcmin radius on the opposing page shows, the only optical objects catalogued within this area are the quasar pair, which is exactly across UGC 9853 from the bright quasar of $z = 1.212$. Accurately along this same line but further to the SE is Mrk 485 with a redshift of $z = .0197$, almost the same as UGC 9853. It is therefore possible that these galaxies are in physical alignment. A blue galaxy (CASG 673) is further along the line to the NW.

Objects Aligned NE-SW from UGC 9853

$z = 1.855$	QSO(BAL)	$V = 18.2$	SBS 1520+530
$z = 1.854$	QSO(BAL)	$V = 18.7$	SBS 1520+530
$z = 1.21$	QSO	$O = 17.1$	CSO 759
$z = .0197$	AGN	$m = 15.0$	Mrk 485

When an attempt was made to find the lensing galaxy it was discovered that it was very close to Quasar B. The space telescope (HST) image is shown here in Fig. 2 (from Faure et al. *A & A* 386, 69). To this possibly biased observer it looks like a typical pair of quasars across the center of a disturbed galaxy. If the magnitudes of the quasars were gravitationally amplified, the closer one would be expected to be much more amplified, yet both quasars have much the same brightness. Perhaps more tellingly, Quasar A is located on the end of luminous material coming out of the galaxy! Immediately around the quasar there may be some trouble with the deconvolution process, but it should be of high priority to get the best high resolution optical image of this critical region.

The spectroscopic redshift of the lensing galaxy (L) is reported as $z = .71$, and its photometric redshift is estimated at $z = .88$. It is interesting to note that there are two absorption line systems in the BAL (broad absorption line quasars), one at $z = .715$ and one at $z = .815$. Of course this does not tell us whether the quasars are involved in the disturbed material of the galaxy or whether they are background objects shining through. The

102

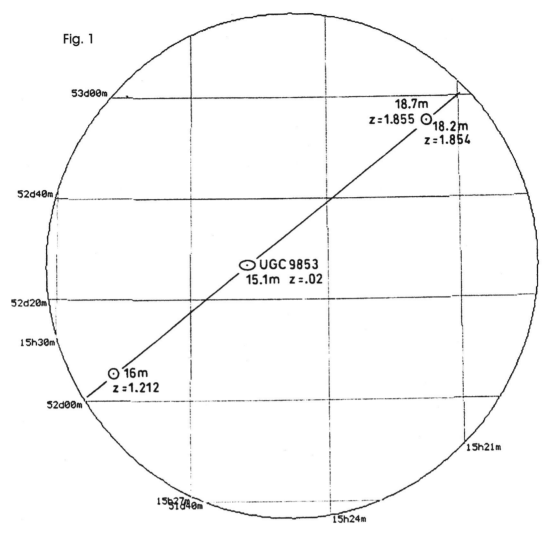

Fig. 1

BAL absorption shortward of C IV probably tells us more about material of intermediate redshift in the lensing galaxy.

UGC 9853 is our proposed origin of both the $z = .715$ galaxy and its $z = 1.855$ quasars. About 2 arcmin north, along its minor axis, it has an apparent cluster of galaxies containing some blue objects. Some brighter blue objects trail away from UGC 9853 to the SE, but any moderately bright AGN's should have been detected by the objective prism, Second Byurakan Survey (SBS).

NEEDED

Deep optical images around UGC 9853, X-ray images of the fields surrounding the components of the association, maximally deconvolved images of quasar pair and the apparently interacting galaxy.

103

Fig. 2

Deconvolved (FWHM = 0.075 arcsec)

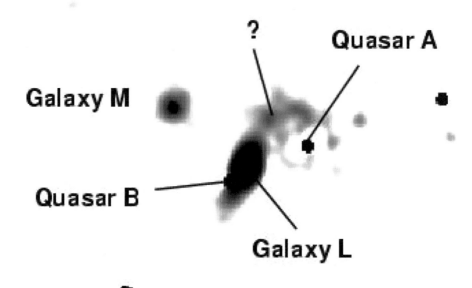

There are two very bright quasars within the pictured field of 60 arcmin radius. The next brightest in the field is R = 18.55 at z = 3.155 and it is close to the bright, active PG quasar with z = .77. The only candidate for the origin of the brightest pair is the compact galaxy MCG+08-29-005, which is of unknown redshift and spectral characteristics. The z = .400 quasar has a number of z = .075 galaxies close around it and a line of X-ray sources across it.

As indicated in the figure, the MCG galaxy is located between MCG galaxies to the N and a ZW Cluster and UGC galaxy to the S. It would be interesting to have their redshifts. It is also interesting to note that there are five high redshift quasars just to the east of the z = .772 PG object, found in the small area, Palomar Grism Survey (1999, *AJ* 117, 40). They are very bright for their redshift, especially the one at z = 3.115. This group is not plotted in the Figure because of the narrow, 8.5 arcmin, strip surveyed.

ALIGNED OBJECTS

z = .400	S1/QSO	V = 16.65	PG 1543+489, 1RX
z = .772	S1/QSO	V = 15.81	PG 1538+477 rad, X-ray

OBJECTS OF INTEREST

z = 3.155	QSO	R = 18.55	Q 1539+4746

plus slightly fainter, high redshift QSO's near the above

NEEDED

Spectra of the central and adjoining galaxies.

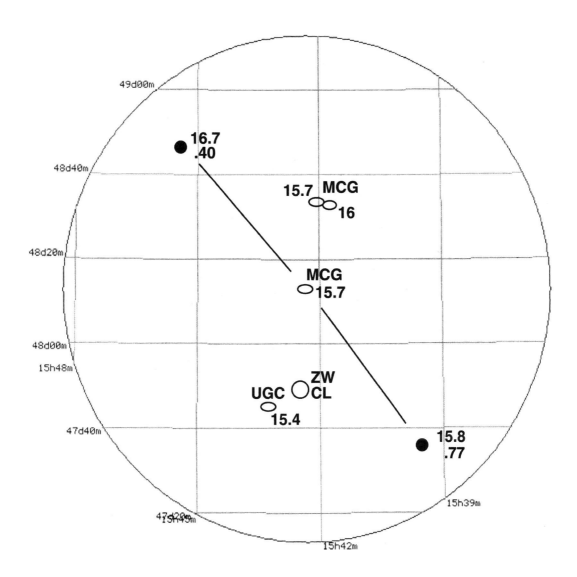

NGC 6140

16^h 20^m 57.5^s +65^d 23^m 24^s
m = 12.6 SBc z = .0026 IRAS source
semi-circled by radio sources

Fig.1: Extension of the line through Mrk 876 leads to a cluster centered on 3C 330. Roughly across 3C 330 are a pair of unusual, clustered objects at z = .63 and .65. X-ray sources in this area are all extended, suggesting their identity as further clusters. Note the Gamma Ray Burster farther out along the line. (Only brightest and most active objects plotted.) Fig.2: PSPC X-ray sources join Mrk 876 with z = .65 cluster and 3C 330 (two crosses to upper right). Fig.3: Radio overlay of HST optical image shows luminous extensions either side of z = .63 galaxy, one of which connects to luminous arc with z = 1.39 (from Jackson et al. A & A 334, L33).

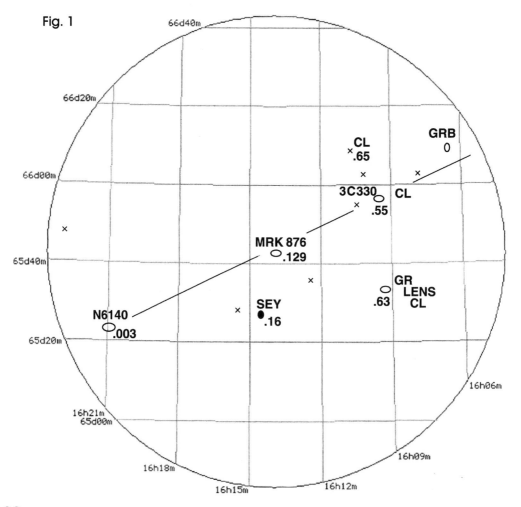

Fig. 1

Fig. 2

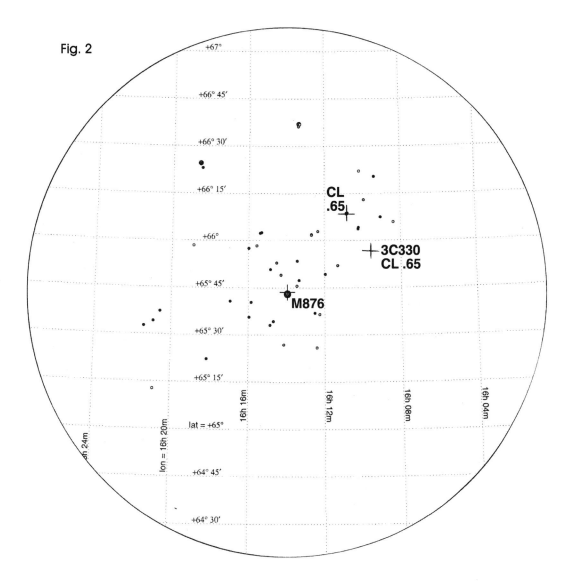

NEEDED

Investigation of numerous additional extended X- ray sources in region.

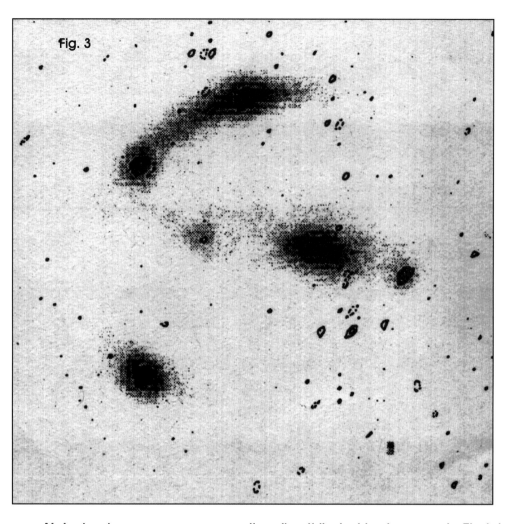

Fig. 3

Note: Luminous arcs are conventionally attributed to degenerate Einstein rings in gravitational lensing. But in Fig. 3 the image obtained with the Hubble Space Telescope (grey scale) shows luminous extensions coming out *radially on either side of the central z = .63 galaxy.* This is the classic ejection pattern (e.g., the optical jets emerging from NGC 1097, as in Plate 2-7 of *Seeing Red*). The optical extensions include radio sources on either side of the central galaxy and the extension to the ENE ends on a luminous arc with a redshift of $z = 1.394$. This lensed "galaxy" is reported to have an A star spectrum and a very high luminosity. This is an unlikely combination. But, in addition, as in many arcs interpreted as lensed background galaxies, the nucleus of the supposed lensed galaxy is at one or both ends of the arc. Another way of putting it is that the brightest part of the galaxy, the nucleus, has somehow gotten lost in the lensing! For more on gravitational lenses see *Seeing Red* (p. 173 ff.).

ALIGNED AGN 's

z = .033 BL LAC	V = 13.78	Mrk 501
z = .034 Sey 2	V = 17.7	

OF INTEREST

z = 1.75	QSO	V = 18.3
z = .692	QSO	V = 20.5
z = .340	QSO	V = 19.2
z = .579	QSO	V = 17.9
z = .151	ACO 2235	m = 17.1

There is a tight cluster of X-ray sources around the famous Mrk 501 (≤ 15 arcmin, Fig. 2). There are strings and lines of sources, one of which extends toward Abell 2235. Such X-ray associations seem to be characteristic of Markarian objects.

Note: Similarity of these AGN redshifts to Mrk objects aligned through M 101 (Mrk 273, 231, 66, 477 and 474 at z = .038, .041, .035, .038, .041). See Part B.

When this field was first investigated little was known about the central galaxy. But just as the present *Catalogue* was being submitted for publication, E. M. Burbidge obtained a spectrum of it. It showed emission lines in the red at z = .016. (An apparent broad blue line at 4000 A is not identified.) It is intriguing to note that when the two AGN's are now transformed to the rest frame of the central galaxy they turn out to have z = .017. This is the ubiquitous 5,000 km/sec redshift found all over the sky and particularly in the Perseus-Pisces filament.

Even though the spectrum needs further investigation, this is again an example of how one additional measurement in a field like this can add immeasurably to the interest and value of the association.

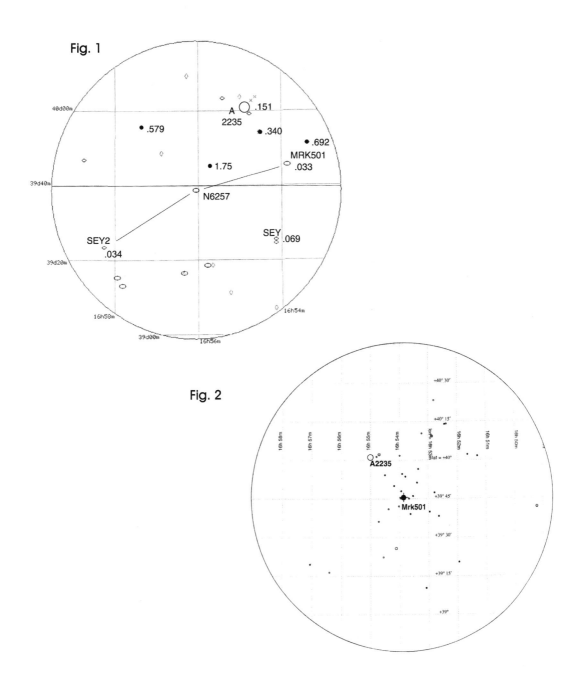

Fig. 1

Fig. 2

Needed

1) Images of NGC galaxy

2) Spectra of X-ray sources near Mrk 501

3) Images and spectra in Abell Cluster

Fig. 1: Five NGC galaxies are within 31 arcmin of this Markarian galaxy, plus quasars of z = 1.14, 1.36, 1.42, 1.47, 1.54, and 3.23. Fig. 2: Inside a radius of only 10 arcmin appears a line of objects, higher redshift than Mrk 892, leading to the strong radio radio quasar 3C 351, which is embedded in a cluster of objects near its own redshift of z = .37. There is a line of X-ray sources leading W from Mrk 892 into the 3C 351 cluster. There are also apparent clumpings of lower redshift galaxies within this concentration. Note the BL Lac object at z = .280 and the AGN on the other side of the cluster at z = .284.

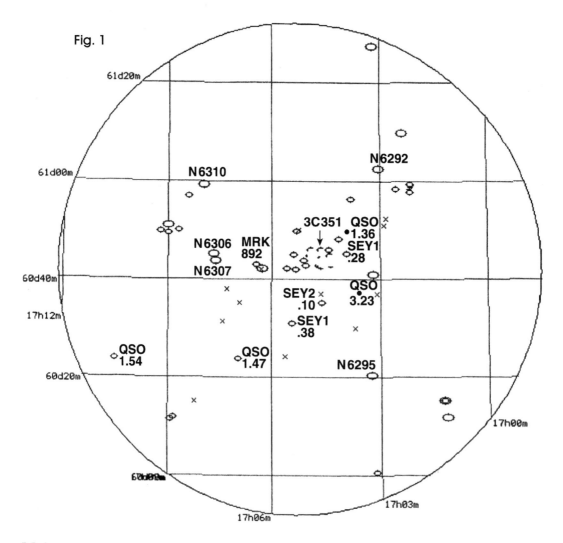

Fig. 1

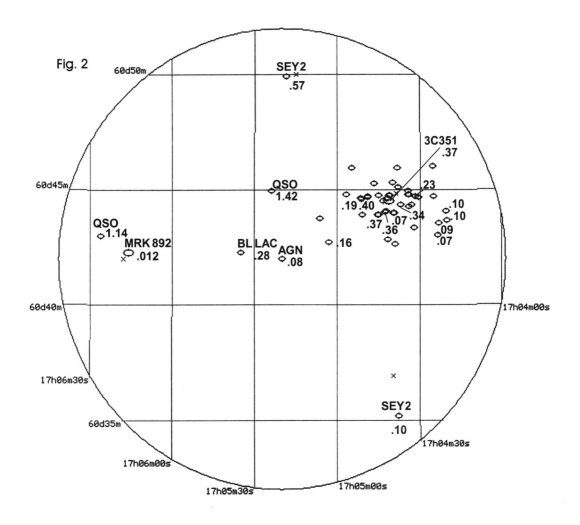

Fig. 2

Objects of Major Interest

z = .012	AGN	m = 15.7	Mrk 892
z = .372	QSR	V = 15.3	3C 351
z = .280	BL Lac	V = 19.4	X-ray
z = .080	AGN	V = 17.7	X-ray
z = .284	AGN	V = 18.9	X-ray

Needed

Identification of additional, numerous X-ray sources in the field. Systematic spectra of galaxies as a function of apparent magnitude. Plots of all galaxies within certain redshift ranges.

One of the close pairs of X-ray clusters in NORAS (Northern ROSAT all sky survey, *ApJS* 129,435) is shown here at z = .32 and .33. In looking for a possible origin of this pair we find a very compact cluster of NGC galaxies about 1.5 degrees away. One of the brighter members of this group is NGC 6965, an active Seyfert 2 Galaxy. In the neighborhood of the two RXC clusters are found quasars of z = 3.80, 3.21, 1.40 and an AGN of z = .06. All of these were discovered in an emission line survey by Peimbert and Costero.

Roughly the same distance on the other side of the NGC galaxies is a group of quasars and AGN's ranging from z = 2.78 to .14. They are mostly from samples of Anderson and Margon plus radio objects. Aside from these compact pairings of unusually compact groups of quasars across the NGC Seyert, the numerical values of the redshifts are intriguing. The Karlsson values of z = .30, .60, 1.41, 1.96 , 2.64 and 3.48 are noticeable, particularly when small corrections are made into the rest frames of the probable parent galaxies.

PAIR OF X-RAY CLUSTERS

z = .333	X-ray cluster	m = ---	RXC J2050.7+0123
z = .321	X-ray cluster	m = ---	RXC J2051.1+0216

ALIGNED HIGH REDSHIFT OBJECTS

z = 3.799	QSO	m = 19.7	
z = 3.214	QSO	m = 20.7	
z = 1.397	QSO	m = 18.5	
z = .060	AGN	m = 21.2	
z = 2.783	QSO	V = 19.1	2038-012
z = 2.260	QSO	V = 20.4	
z = 2.3	QSO	V = 19.9	
z = 2.100	QSO	m = ---	X-ray quasar
z = .609	QSO	V = 18.39	MS 20373-0035
z = .142	S1	V = 18.60	MS 20395-0107

NEEDED

Spectra of objects along the line between the two high redshift groups.

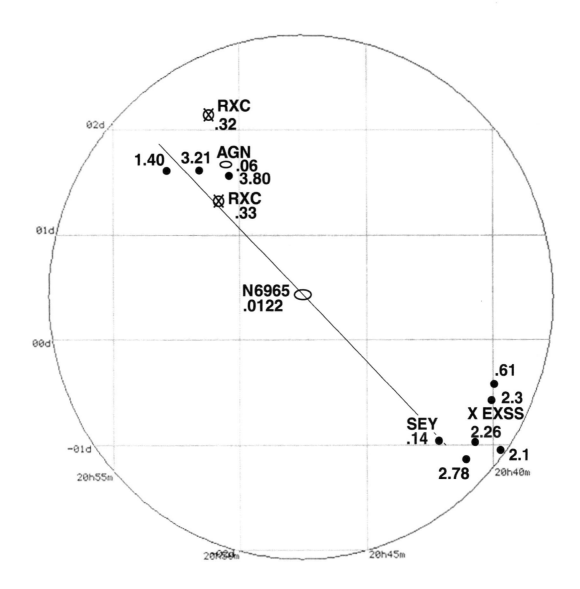

MCG-04-51-007

(AM 2137-225)

$21^h\ 39^m\ 59.3^s$ $-22^d\ 41^m\ 32^s$

m = 15.0 z = .030

also MS 2137.3-2353

Fig.1: At about the center of this elongated cluster lies this moderately bright z = .030 galaxy. A line of fainter galaxies extends NW from it with redshifts z = .031 to .036. The cluster AS 963 (Abell southern extension) includes NGC 7103, N7104 and two IC galaxies (redshifts unknown). Another cluster, S85 426 (Shectman) lies a little further out.

Fig.2: In the opposite direction A 2357 lies along the line with z = .123. Note the galaxy further out along this line with z = .125. Note also the three galaxies with z = .055 to .057. There are many PHL BSO candidates in the area. The shaded area represents the region which is confused by the globular cluster NGC 7099.

Objects of Major Interest

z = .646	QSO	V = 17.0	X-ray
z = .313	ClGal	V = 18.5	X-ray
z = .123	ClGal	m = 17.0	A 2357
z = .81	RadGal	m = --	Radio
z = .033	ClGal	m = --	AS 963
z = .--	ClGal	V = --	S85 426

Needed

Redshifts of galaxies SW of MCG-04-51-007, redshifts within, and deep images of clusters.

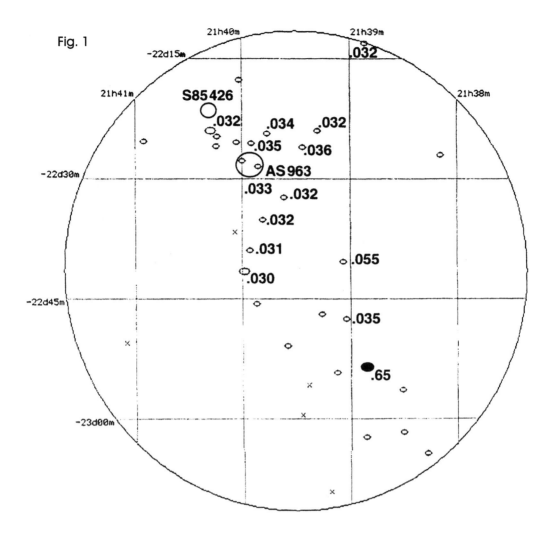

Fig. 1

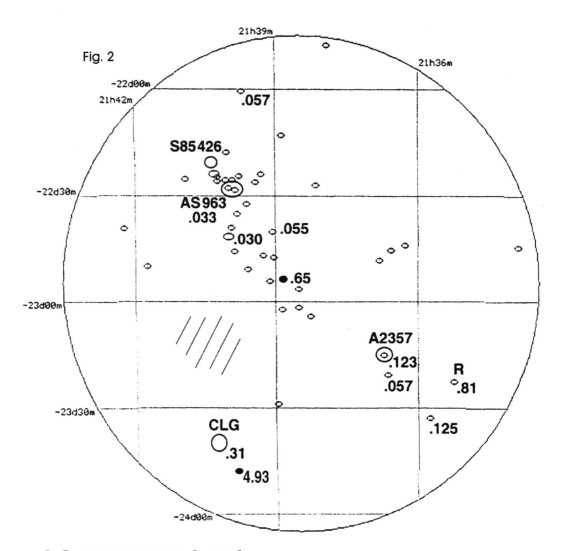

Fig. 2

A Gravitational Lens?

About 43' S of the z = .030 central galaxy in Fig. 2 is a very strong X-ray cluster of galaxies at z = .313. Fig. 3 here shows the Hubble Space Telescope image of the central galaxy (Sand, Treu and Ellis 2002, *ApJ* 574, L129). The picture features a supposed double gravitational arc just above it and a supposed *radial* (!!) arc emerging from this active central galaxy. It has been argued elsewhere (*Seeing Red*, Chap 7 and Fig. 7-10) that so called gravitational arcs in galaxy clusters are actually due to material ejected outwards meeting previously ejected shells. In this particular case of MS 2137.3-2353 the high resolution HST image in Fig. 3 confirms by simple inspection that the feature emerging directly toward the tangential arc is a high surface brightness, linear jet. As for tangential arcs, it has been obvious since the beginning of this subject that they

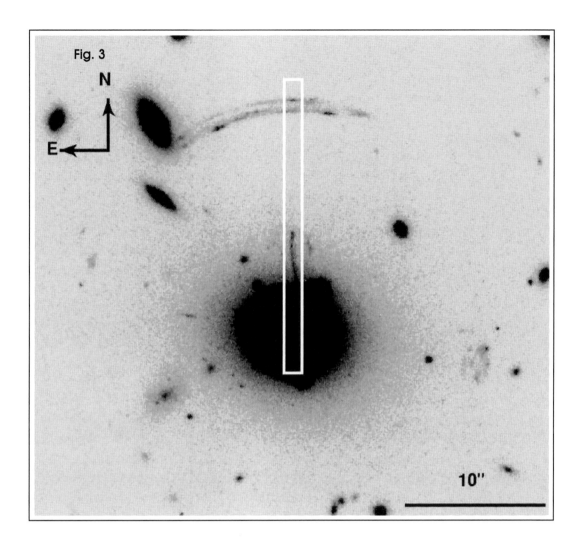

Fig. 3

show no nucleus—as they must show if they are the lensed images of a background galaxy.

What is new in this MS object is that the arcs have been measured at $z = 1.501$ and the jet at $z = 1.502$. Obviously this is the same material ejected from the center of the active galaxy and not the fortuitous arrangement of identical redshift background galaxies. As a finishing touch we note a newly discovered quasar only about 11' away with $z = 4.93$ (Fig. 2). Transformed to the rest frame of the central galaxy (which is already close to the quantized redshift peak of $z = .30$) this quasar then has a redshift of $z_0 = 3.52$. The quantized peak for the latter is $z_p = 3.48$ where the adjoining peaks are at $z_p = 2.64$ and 4.51.

Two very high redshift (and hence rare) QSO's are closely aligned with, and only 7.9 arcmin distant from the peculiar galaxy NGC 7107. A companion-like object NW in the disk of the galaxy has exactly +72 km/sec relative redshift. Within 20 arcmin of NGC 7107 there are two quasars of z = 1.64 and 2.38 to the NNE and four quasars of z = 1.39, 2.01, 2.21 and 2.16 to the SSW. The quasars were identified in UKSTU fields monitored for variability.

ALIGNED OBJECTS

z = 3.06	QSO	V = 20.2	Q2139-4504A
z = 3.25	QSO	V = 20.3	Q2139-4504B
z = .16	GAL	m = --	
z = .00749	GAL	m = --	

Note: In a large region around here are numbers of bright NGC galaxies. About 30 arcmin N of NGC 7107 lies the "Francis filament" of galaxies with z = 2.38 and two quasars of z = 3.22 and 3.23. There are also groups and filaments of galaxies from the Las Campanas redshift survey and Abell 3800. The z = 2.38 objects are uncannily close in redshift to the cluster of z = 2.38 QSO/AGN's discovered by Windhorst and Keel (see *ApJ* 525, 594)

NEEDED

Deeper, higher resolution images of NGC 7107. More extensive study of redshift content of filaments.

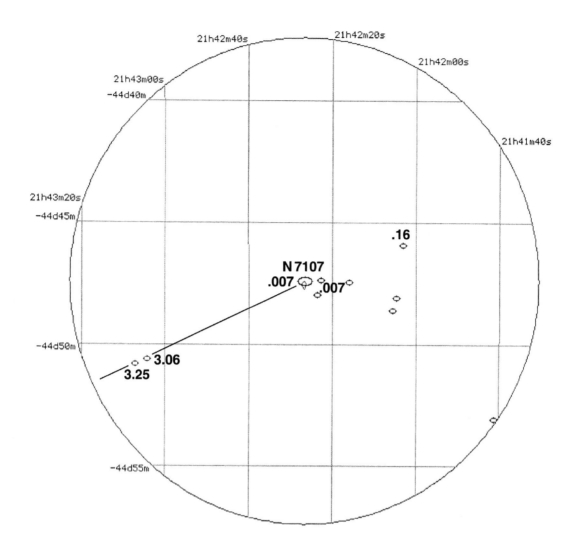

The galaxy cluster Abell 2361 (extending into 2362 and 2360) is strikingly elongated (see included Simbad map, Fig. 1). Following the direction of the elongation leads to NGC 7131. This same direction leads to another galaxy cluster, Abell 2400, exactly the same distance on the other side of the central galaxy. It is also elongated along this same direction. There is even a kink in the elongation which accurately matches the same feature in the NGC 2361 galaxy cluster. The whole configuration suggests a schematic of rotating arms in a barred spiral. In this case, however, they would be galaxy forming ejections instead of star forming ejections.

At the beginning of the kink in the A 2400 cluster there is a galaxy of z = .018, establishing a link with NGC 7131. In the A 2361 elongated complex, a sampling of galaxies shows z = .05 to .08, which is too much spread in velocity for a stable cluster, particularly one of such non-equilibrium shape.

The brightest galaxy in the field is NGC 7171 at B = 13.0, z = .009. There is a O = 18.6, z = .73 quasar very near it. This galaxy could be the oldest in the field, with NGC 7131 being the younger, more active source of ejection. There is a PKS quasar with O = 19.5, z = 1.80 immediately to the NE of N7131 on the path to A 2400.

See Appendix B at end, Ejection Fig. 14, for a different presentation of this cluster configuration.

OBJECTS OF MAJOR INTEREST

z = .061	C(luster)	m = 16.7	A 2361
z = .088	C	m = 16.5	A 2400
z = .018	gal	m = 14	NGC 7131
z = 1.80	QSR	O = 19.5	near N7131
z = .009	Sbl	B = 13.0	NGC 7171
z = .73	QSO	O = 18.6	near N7171

NEEDED

Systematic spectra of galaxies as a function of apparent magnitude. Plots of all galaxies within certain redshift ranges should help clarify the relation of the central galaxy to these two extremely elongated clusters.

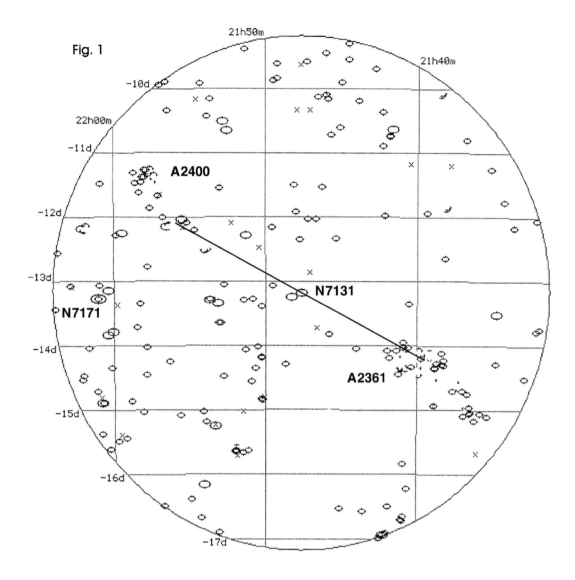

Fig. 1

Figures 1 and 2 show expanded plots of the clusters on either side of NGC 7131. The galaxies are so dense that even on an expanded scale some of the symbols do not plot completely. See Appendix B, Fig. 14 for a representation of galaxy density which best shows the overall outline of the clusters.

Fig. 2

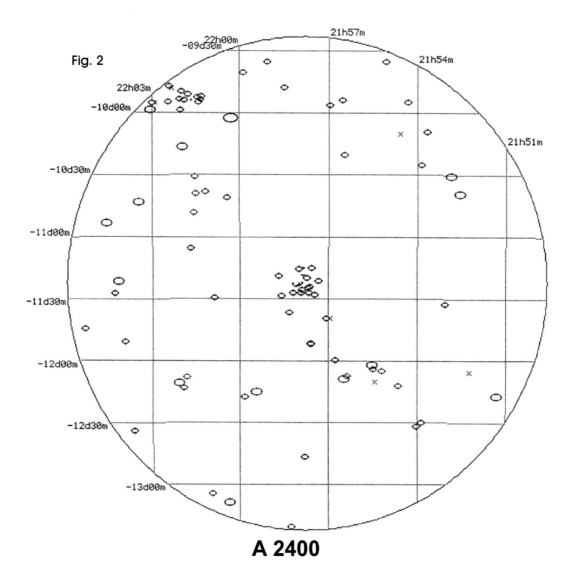

A 2400

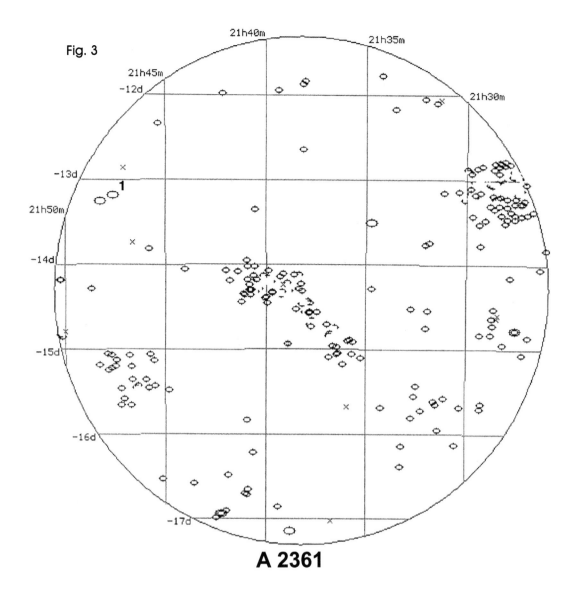

A 2361

Fig. 3

The Abell Cluster 2464 has a surprisingly large redshift z = .56. Within it, a proposed gravitational "arc" of z = 1.116 was discovered and later an apparently associated "arc" of z = 1.334. The latter two objects would seem more naturally ejected from cluster galaxy C, but the reader should consult the images shown in *A & A* 290, L25, 1994.

The question asked here is whether there is any evidence for the ejection of this unusual cluster from a nearby, low redshift object. The accompanying map shows that the supposed lens indeed lies in a narrow string of objects extending about 1.4 deg. to the NE of NGC 7351. This line consists of radio sources and mostly associated clusters of galaxies and some BSO candidates for quasars.

Starting from the NW, there is a cluster of galaxies elongated along a line at about p.a. 130 deg., or accurately back toward NGC 7351. Next there is the survey X-ray source, 1RXS J223732.3-033805 which has been measured by Y. Chu to be a quasar of z = .35. Then there is the survey X-ray source 1RXS J223826.3-034356 which lies between 2 BSO's of R = 18.8 and 18.1 mag. The next galaxy inward on this line was measured to have a redshift z = .06. An unidentified emission line at about 8771 Angstroms was also present in this spectrum and needs to be confirmed and identified. Next comes the Abell cluster at z = .56 which is supposed to be lensing the apparently connected higher redshift objects. A short distance SE of this cluster is a radio source which has apparently been measured twice as PMN J2239-0402 and Cul 2236-042. It lies close to two BSO's of R = 18.2 and 18.9 mag. Next comes the radio source 4C-04.85 which appears to be a Blue compact galaxy in a rich cluster (Bcg). Finally near NGC 7351 is a PHL BSO (at 22/41/04.8, −04/18/10) and another small elongated cluster. The latter, as well as other clusters along this line, are visible on Digital Sky Survey (dss) downloads.

NEEDED

Verification of clusters along the line.

Redshifts and deep images.

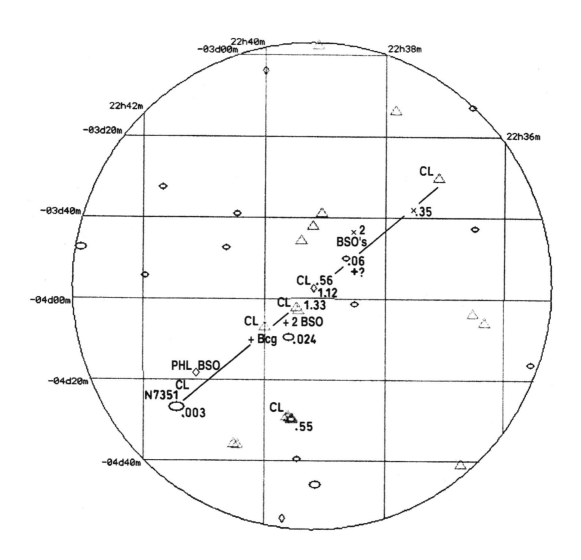

OBJECTS ALIGNED NW-SE TOWARD NGC 7351

z = .--	RadCl	m = ---	PMN J2237-030
z = .35	QSO	V = 16.3	1RXS
z = .--	BSO	R = 18.8, 18.1	1RXS, 2 BSO's?
z = .06	GAL	m = 16:	DBL z?
z = .56	C(luster)	m = 17.8	A(bell) 2464
z = 1.12	grav. lens	m = ?	Cl 2236-04
z = .--	BSO	R = 18.2, 18.9	PMN, CUL in Cl
z = .--	RadCl	m = ---	4C-04.85
z = .024	GAL	m = 14	NGC 7344
z = .--	BSO	R = 16.8	PHL 372
z = .--	Cluster	m = ---	sm cl NNW of N7351

129

The BL Lac object shown in this figure was discovered as a hard ROSAT X-ray source. The only bright, nearby galaxy is NGC 7378, which turns out to be a Seyfert 2. Surprisingly, an almost equally active radio quasar of z = .630 lies on a line back to NGC 7378 from the BL Lac. The latter quasar is radio and X-ray bright and is highly polarized—all marks of a BL Lac type object. The two aligned objects therefore show a similar type of activity. (The apparent magnitude of the z = .226 BL Lac is uncertain, Simbad listing it as 15.9 and NED as 20.0. The discrepancy may be caused by an error or the nebulous character of the image or variability). There is a PKS quasar S of NGC 7378 and an AGN to the W (z = .117 not shown) which may also originate in the Seyfert 2 NGC galaxy.

ALIGNED OBJECTS

z = .226	BLLac	m = ?	RBS 1888
z = .630	QSO	V = 16.45	rad, X-ray source

OF INTEREST

z = 1.892	QSR	O = 19.1	J 224752.6-123720

NEEDED

Deeper, higher resolution images of central galaxy. More extensive search for quasars in the field and objects along the line.

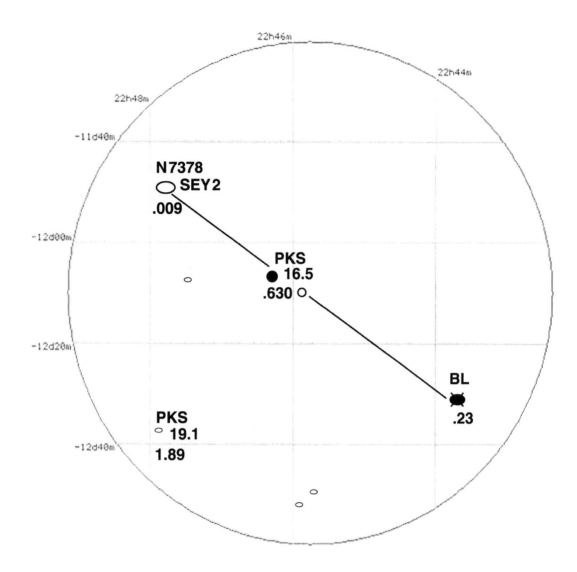

A pair of Sloan Digital Survey (SDSS) quasars at z = 3.64 and 3.68 fall quite close to each other in the sky. Well centered between them is a UGC spiral of z = .025. Also striking is a z = 2.65 quasar, previously discovered, which falls in the same alignment. Perhaps most interesting of all are the redshifts of these three quasars. When transformed into the rest frame of the central galaxy they come out z = 3.53, 2.56 and 3.57. The quantized redshifts in this range are z = 3.48 and z = 2.64. In turn this means the observed redshifts fall within cz = +.01c, −.02c and +.02c of the predicted redshifts, implying quite small radial velocity components.

Note the two gamma ray bursters (GRB's) apparently belonging to the association.

ALIGNED OBJECTS

z = 3.64	QSO	m = --	SDSS J230639.65+010855.2
z = 2.647	QSO	V = 22.45	J230414.6+003739
z = 3.68	QSO	m = --	SDSS J230323.77+001615.2
z = ----	BSO	V = 16.6	X-ray source (see below)
z = ----	GRB	m = --	GRB 991217, 230315+0015
z = ----	GRB	m = --	GRB 00519, 230439+0110

NEEDED

Deeper images of central galaxy. More extensive search for quasars in the field and objects along the line. An X-ray source between the z = 2.65 and 3.68 quasar is close (30 arcsec) from a BSO (R,B = 16.6, 16.6) at 23/03/41.2+00/27/09.4.

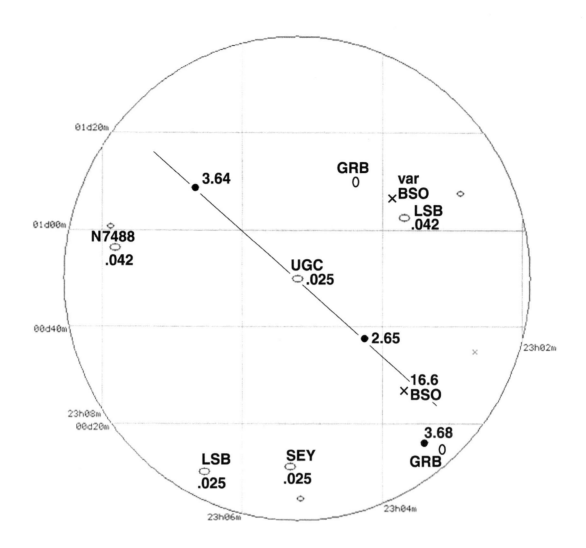

The plotting program by Chris Fulton indicated a concentration of quasar redshifts with Karlsson values at a position around 23h12m–28d10m. Viewing these 2dF quasars as listed by Simbad shows the elongated grouping of quasars pictured in here in Fig. 1. It is apparent that this rather dense grouping forms a cone-like distribution having at its apex the very bright NGC 7507 and the companion galaxies that make up its group (Arp/Madore 2309-284). In Fig. 1 the observed redshifts are written next to each quasar, filled circles for redshifts falling close to the ...30, .60, .96, 1.42, 1.96, 2.64, 3.48... quantized peaks and open circles for redshifts falling between these peaks. It is clear that the quasars stretching up and to the NW from NGC 7507 are almost all near the quantized peak values. (Since the galaxy of origin, NGC 7507, is such low redshift no correction has been applied to the observed redshifts, i.e., $z_0 = z$).

In order to make a pure test of the quantization in this 35 arcmin radius region, without reference to the presumed galaxy of origin, the histogram in Fig. 2 has been constructed. Each quasar redshift has been differenced with its nearest quantized peak ($\Delta z = z_{peak} - z_0$). This number is divided by half the interval to the next nearest peak, yielding the probability of a random point falling this close to the peak. If the redshifts are distributed randomly with respect to the peaks then we would expect a flat distribution of points in Fig. 2. In fact there is a strong excess of redshift values much closer to the peaks than would be expected by chance. A simple estimate can be made that the uncorrelated points in Fig. 2 number $2 \times 6 = 12$ leaving 13 redshifts with $P_{av} \leq .25$ in excess. The probability is about $P_{tot} = .50^{13} = 1.2 \times 10^{-4}$.

As extreme as this result is, the actual association is even more significant because it is possible to draw a cone shaped perimeter, with NGC 7507 at one end, which contains 15 quasars, *only one of which does not fall near a Karlsson peak*. It seems that this would strongly support the ejection origin for the excess number of quasars from this very low redshift, bright galaxy.

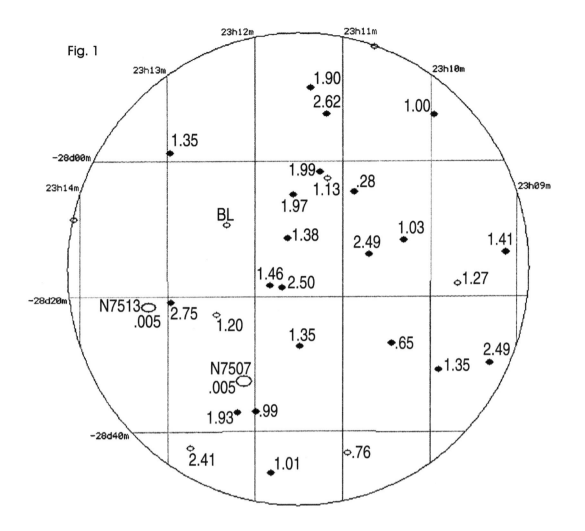

Fig. 1

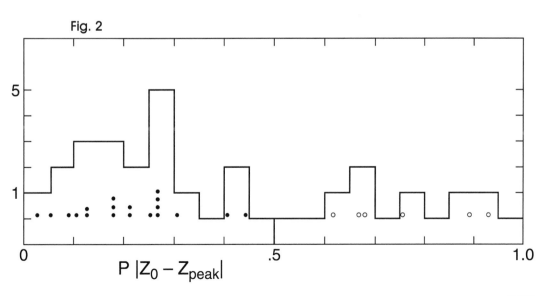

Fig. 2

$$P |Z_0 - Z_{peak}|$$

This is a nearby *star burst galaxy*, surrounded by radio sources, with a blue jet/arm which has an early type spectrum.

Fig. 2 shows a Palomar 200-inch photograph of NGC 7541 which is aligned between two very strong radio sources (from *Astrofizika* 4, 59, 1968). The blue jet/arm is an important feature because it may represent that brief period when the track of an ejected quasar is being turned into a spiral arm by the rotation of the disk of the galaxy. The high surface brightness, early spectrum of this feature would suggest intense recent star formation along this ejection track.

The two very strong radio sources across NGC 7541 have been known since 1968. Only recently, however, has the second one, 3C 458, been measured to be a quasar at z = .29. As a consequence we now have a close pair of strong radio quasars across the active galaxy seen in Fig. 1. This almost exactly duplicates the configuration across Arp 227. (The latter was discussed in the introductory section, Intro Fig. 1 and 3, and calculated to have a probability of accidental occurrence of 2×10^{-9}.)

Fig. 3 shows a larger area around NGC 7541, where Abell galaxy clusters are aligned in roughly the same direction as the 3C quasars. There is also evidence for ejection of quasars perpendicular to this direction. (Note the correspondence between quasar redshifts and Karlssson formula peaks, z = .30, .60, .96, 1.41 and 1.96.)

Perhaps the most crucial result, however, is shown in the last map (Fig.4). The Abell cluster 2552 is believed to be centered on an X-ray, radio AGN of z = .30 (although there are some brighter galaxies at z = .13). *But the X-ray sources in this cluster, in the 1 RXH reduction, are distributed linearly and accurately back to NGC 7541!*

In Appendix B also see Figs 3 and 4 and discussion therein.

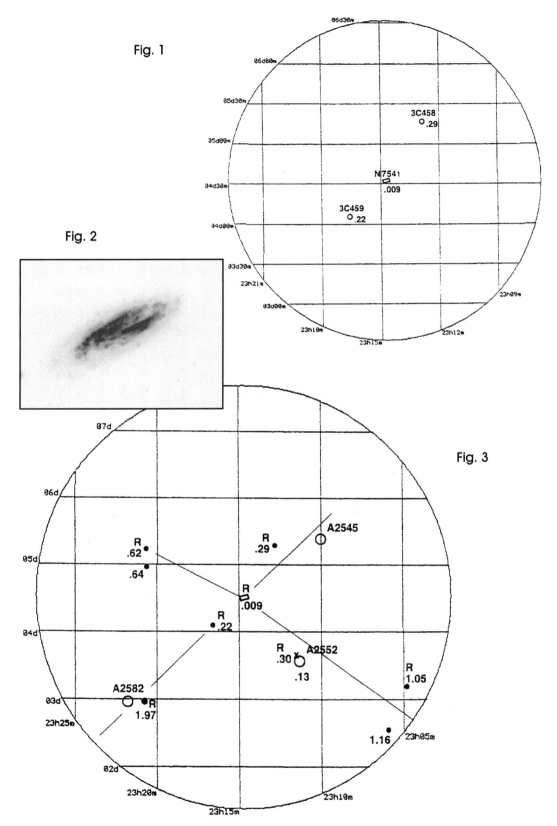

Fig. 1

Fig. 2

Fig. 3

137

As nearly as can be determined a galaxy in this cluster, Abell 2552, was measured at redshifts of z = .128 and.137 (Struble and Rood 1999). There is a dominant X-ray galaxy at the center of the cluster, however, that causes the cluster to be classified as a strong X-ray cluster at z = .300 (Böhringer et al. 2000; Ebeling et al. 2000). In the interim the z = .133 galaxy was reported as having a peculiar emission line spectrum, peculiar morphology and judged too bright to be the dominant galaxy for the surroundingcluster (Crawford et al. 1999)! What appears to this writer is that active, and perhaps interacting, galaxies of much different redshift have been ejected from the parent galaxy NGC 7541.

Cases of ejection of different redshift objects in the same line are prevalent—e.g., the PG 1244+026 example discussed in Appendix B, the z = .008 and .81 objects only 2.4 arcsec apart in the line from NGC 5985 (A&A 341, L5), the z = .34 and .75 objects, which are separated by only .25 arcsec in 3C 343.1 (Intro Fig. 11) and the various members of Stephan's Quintet associated with NGC 7331 (e.g., as revealed by various observations discussed in *Seeing Red*). The cluster shown in Fig. 4 would seem like another elongated cluster composed of galaxies with different redshifts pointing back to their active galaxy of origin.

ALIGNED STRONG RADIO QUASARS

| z = .22 | QSR | V = 16.7 | 3C459 |
| z = .29 | QSR | m = 20.0 | 3C458 |

OUTER ALIGNMENTS

z = 1.97	QSR	V = 18.9	R(adio)
z = .62	QSO	O = 19.0	R
z = .64	AGN	R = 20.4	
z = 1.05	QSO	V = 15.8	R
z = .30	AGN	m = -----	R + X (Cluster)
z = .13 ?	C(luster)	m = 18.0	A(bell) 2552
z = --	C	m = 17.6	A 2545
z = --	C	m = 17.6	A 2582

NEEDED

More data on galaxy jet/arm and clusters.

Fig. 4

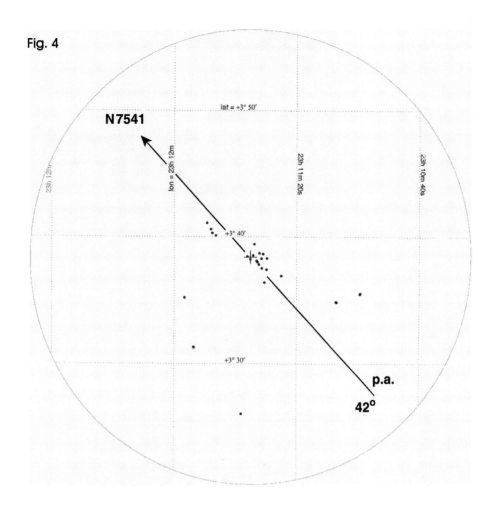

N7541

lat = +3° 50′

23h 12m

23h 11m 20s

23h 10m 40s

lon = 23h 12m

+3° 40′

p.a.

42°

+3° 30′

139

On the facing page, active and unusual objects are plotted in a field radius of 1.9 degrees around this active Seyfert/Quasar. There are a pair of bright X-ray quasars diametrically across it at about 20′ and 40′ with z = .72 and 1.35. There are also two BL Lac objects and a ULIRG to the SE. A pair of Abell clusters brackets the Seyfert in approximately the same direction.

A group of X-ray sources forms a rough arc between the Seyfert and Abell 2659 (small dots in the Figure). Near the center of these X-ray sources some very red objects have been recorded by HST. The 12 X-ray sources are from a ROSAT HRI exposure on the galactic variable R Aquarii. They do not show on the PSPC image.

ALIGNED OBJECTS

z = .137	Sey	r = 15.4	MS 23409-1511
z = .718	QSO	r = 17.4	RXJ 23408-1426
z = 1.347	QSO	r = 17.7	RXJ 23447-1500
z = .224	BL Lac	V = 19.2	MS 2343-151
z = .--	BL,HP	V = 19.22	MS 23427-1531
z = .241	ULIRG	m = 19.26	IRAS F23451-1538
z = .--	GCl	m$_{10}$ = 17.6	Abell 2648
z = .--	GCl	m$_{10}$ = 17.0	Abell 2659

NEEDED

I.D. of X-ray sources near ACO 2659 and Sey.

Note: As discussed in the Introduction, quasars associated with galaxies of appreciable redshift need to have their redshifts transformed into the redshift frame of the parent galaxy. In the present case, (1 + .72)/(1 + .14) = 1 + .51 and (1 + 1.35)/(1 + .14) = 1 + 1.06. The deviations then from the quantized, intrinsic values of z = .60 and .96 are cz = −.06 and +.05. These values ostensibly represent the ejection velocities, cz = −18,000 km/sec toward the observer and +15,000 km/sec away from the position of the central galaxy. It is very encouraging to note that these are very close to the ejection velocities calculated from the best pairs in the Introduction (section d, ejection velocities) which did not have

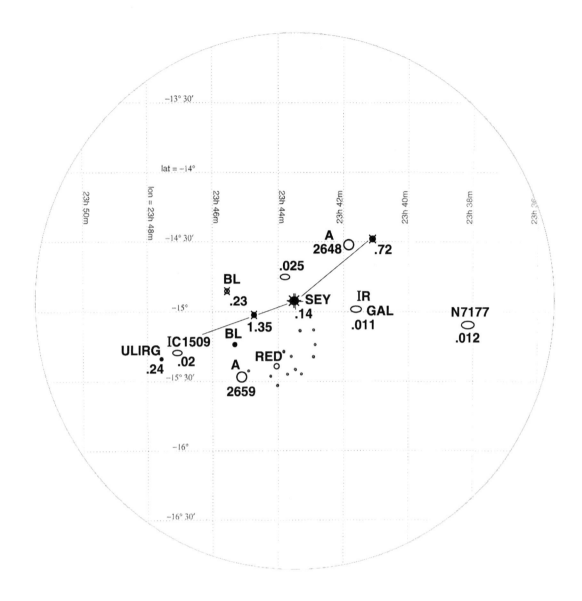

appreciable corrections for the redshift of the central galaxy. See also quasars in NGC 622 field discussed earlier in the *Catalogue*.

These corrections should be made throughout the associations pictured in this *Catalogue*. For example, in the pair across A 27.01 (ESO 359-G019) the values come out $\Delta z = -.05$ and $+.11$.

Hickson compact Group No. 97 consists of 5 galaxies of mixed type with mags. B = 14.2 to 16.3 mag. The redshift of the brightest is z = .023. The X-ray map here shows sources extending to the NE in the direction of PKS 2349-014, a very bright radio quasar (z = .17). *The X-ray sources around the quasar are conspicuously extended, closely along a line back toward HCG 97.* In exactly the opposite direction and distance from HCG 97 we find the very bright radio quasar PKS2340-036 (z = .896). Slightly further along this line we find the linked, double galaxy Arp 295 with z = .023.

Suggested interpretation: This group of IC galaxies was previously a larger, brighter galaxy which broke apart in the ejection of the higher redshift material. Some of the original material (z = .023) was carried along. A number of z = .02 galaxies are in the vicinity—and HCG 97 (and associated Shakbazian very compact objects) may have originally been part of a chain including NGC 7738 and NGC 7739 (to the NNW, just out of the frame).

To the SW is seen the edge of an Abell cluster ACO 2656 containing an RXJ source at z = .079. Further to the SW in this direction is an IRAS galaxy with z = .080 and 48 arc min SW of the cluster along the same line is a 1RXS source which is a 17.2 mag Seyfert 1 of z = .076. These are also candidates for ejection from HCG 97, along with ACO 2644 to the NW at z = .070, a cluster which also appears to have higher redshifts associated with it.

ALIGNED STRONG RADIO QUASARS

z = .174	QSR	V = 15.33	PKS 2349-01
z = .896	QSR	V = 16.02	PKS 2340-036

OUTER ALIGNMENTS

z = .023	dbl gal	m = 14.5	Arp 295
z = .993	QSO	V = 16.91	UM 186

The quasar UM 186 at z = .993 is very near PKS 2349-01 and forms a similar redshift pair with PKS 2340-036 at z = .896. The mean is very close to the redshift peak of z = .96.

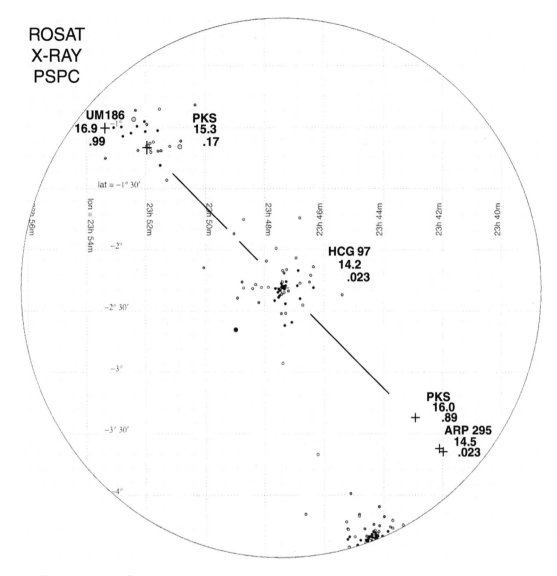

ROSAT
X-RAY
PSPC

UM186
16.9
.99

PKS
15.3
.17

lat = −1° 30′

lon = 23h 54m

23h 56m

23h 52m

23h 50m

23h 48m

23h 46m

23h 44m

23h 42m

23h 40m

−2°

HCG 97
14.2
.023

−2° 30′

−3°

−3° 30′

PKS
16.0
.89

ARP 295
14.5
.023

−4°

POSSIBLE ASSOCIATION

z = .079	C(luster)	m = 16.2	A(bell) 2656
z = .070	C	m = 16.6	A 2644
z = --	C	m = 18.0	A 2631

The bright quasar PKS 2349-01 appears to be a very active object. UK Schmidt objective prism and CFHT slitless spectrograph (or "grens") identifications in this area record at least 10 quasars from z = 3.0 to z = .45 distributed within 30 arcmin along an axis through it at about p.a. = 355 deg. This would be a rich field for further investigation.

Appendix A

The following two Appendices were originally written as Journal papers. But when the *Catalogue* turned out to introduce so many associations of different kinds of objects, it was felt that going into detail in some particularly key cases could illustrate their possible relation to each other and also give an example of the potential information contained in the *Catalogue* associations. The following advantages to presenting this material together in book form were then envisaged:

1) A large amount of material bearing on the subject of associations of different redshift objects could be published together instead of being spread out through a voluminous literature, which is for the most part based on antithetical assumptions.

2) The struggle to overcome inevitable resistance by referees and editors could be avoided. The evidence could be presented uncensored and discussed more frankly. Of course readers have to be their own referees.

3) People who were interested in this particular kind of evidence could obtain the book and use it as a condensed working reference.

4) A large enough body of data would be in one place so physical interpretation could be attempted and relation to theory initiated.

5) Since the *Catalogue* is only a sample of many associations yet to be found, readers would be better able, with this added discussion, to select the kinds of further associations that promise the most progress.

THE OPULENT NEIGHBORHOOD OF THE NEARBY GALAXY M 101

INTRODUCTION

As an example of how one of the associations in the main *Catalogue* might be analyzed from many different aspects, the ScI galaxy, M 101 is treated here as a separate investigation. Of course M 101 is one of the nearest galaxies that can be examined, so we are able to deal with brighter apparent magnitude objects than in most fields in the *Catalogue*. Further, because of its closeness, 6.7 Mpc, (or perhaps even as close as 3.6 Mpc) an extended area needs to be studied. In the end we will see strings of galaxies, galaxy clusters, bright Markarian and 3C radio objects and bright quasars extending from the central galaxy out to over 10 degrees radius on the sky. These alignments appear to point to past, recurrent ejection of high redshift objects. It is suggested that these proto-galaxies have evolved to lower redshift with time, giving rise to linear distributions of galaxies of similar redshifts, which are nevertheless still higher today than the low redshift of their galaxy of origin.

It will be seen that galaxy clusters paired and aligned across M 101 extend along lines back to M 101. This supports previous evidence on associations of higher redshift galaxy clusters with low redshift galaxies. In conjunction with similar evidence for high redshift quasars, this suggests that high redshift ejecta can fission into smaller pieces and evolve into clusters of lower redshift galaxies.

The findings in the M 101 region here are bolstered by similar examples from the *Catalogue* where filaments of galaxies stretch away on either side of the central galaxy, ending in elongated, higher redshift clusters. Of course, placing clusters so nearby yields lower luminosities for the cluster galaxies—in contrast to the unreasonably high luminosities calculated for cluster galaxies on the usual redshift-distance law. Larger galaxy cluster distances calculated from the Sunyaev-Zeldovich effect are briefly considered.

Genesis of the M 101 Investigation

A recent investigation (Arp and Russell 2001) reported a number of cases of galaxy clusters paired across large, nearby galaxies. The reality of these physical associations was reinforced by similarities in the redshifts of paired clusters, the distribution of X-ray and radio quasars along the directions to the clusters, in many cases the activity of the central galaxy, and in some cases elongation of the clusters back toward the central galaxies.

One of the cases presented in the above reference was that of two bright clusters of $z = .070$ and $z = .071$ aligned diameterically across the Scl spiral, M 101. (They are more exactly aligned and remarkably well centered when plotted in the polar coordinates of Fig. 6.) I had known that Mrk 273, regarded as one of the extreme "ultra luminous" infrared galaxies, was relatively close to M 101 on the side toward the $z = .070$ galaxy cluster (Fig. 1). I then happened to notice the Hickson Compact Group No. 66 at $z = .070$. It too turned out to be in the direction of M 101 from the cluster, with $z = .070$. While I was considering this development, an astronomer who has extensive knowledge of the field, Amy Acheson, sent me a question: "Why do the famous, active objects like 3C 295 tend to fall close to bright galaxies?" I plotted the position of 3C 295, as in Fig. 1 here, and therepon decided that I had to investigate in detail all the various kinds of objects around this particularly large, nearby galaxy.

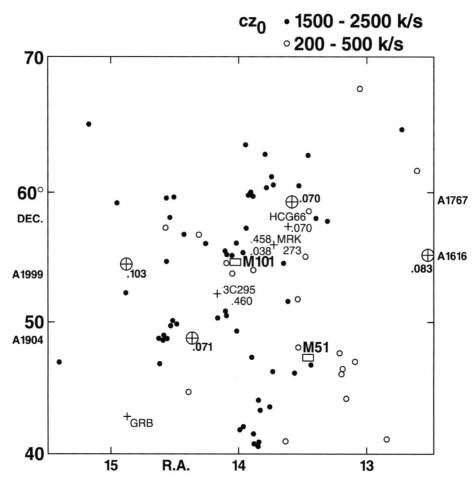

Fig. 1 - Companions around the nearby galaxy, M101 (200-500 km/sec) are plotted over a wide area. M101 has a redshift of 251 km/sec, but galaxies of redshift 1500-2500 km/sec also appear concentrated around it, many in a predominantly NW–SE line. The circled plus signs represent all Abell clusters with m_{10} brighter than 16.3 mag. The clusters which pair NW-SE across M101 have mean redshifts of $z = .070$ and $.071$. The plus signs designate special objects described in text. The plus symbol SE from M101 marks the brightest gamma ray burst in the sky reported so far.

THE EXTENDED REGION AROUND M 101

The filled circles in Fig. 1 plot of all galaxies between $cz = 1500$ and 2500 km/sec over a large area around M 101. They are very sparse on the edges of the region but increase steadily toward the center, to the position where M 101 is located. It is difficult to see how this concentration toward M 101 could be a selection effect, since galaxies of both higher

148

and lower redshift are scattered more or less uniformly throughout the area. (See Arp 1990 for details of the original analysis.) Moreover, these galaxies form lines radiating from the position of M 101 outwards. The line radiating NW-SE from M 101, however, is the most populated. The open circles show low redshift companions, near the $z = .0008$ of M 101, reaching out into the same areas—demonstrating that physical association with M 101 reaches out to this distance.

The latest update in Fig. 1, shown here, has marked the positions of the brightest Abell Clusters (as in the earlier Fig. 7 of Arp and Russell 2001). But also added are three plus signs showing the positions of Mrk 273, HCG 66 and 3C 295. These three objects, along with two bright Abell clusters, fall roughly in the filament of galaxies which extend in the NW–SE direction which was identified in 1990.

Mrk 273 is one of the three major, ultra luminous infrared galaxies known (Arp 2001). It is remarkable that it falls this close to, and with this orientation with respect to M 101. It is a strong X-ray and radio source and in this respect is similar to higher redshift, active objects ejected from more distant central galaxies. In the present case, however, the apparent brightness of both the central galaxy and the AGN implies a lesser distance and accounts for the larger apparent separation (~3 deg.).

Further along in this direction we encounter HCG 66 (Hickson 1994). This is a compact chain of six galaxies, four of which have redshifts which average to $z = .070$. As a group it is like a small cluster, or sub-cluster (albeit in a non equilibrium configuration). It would seem to be related to Abell 1767 just to the NNW, which has an identical redshift. In this respect it would seem to be an extension of Abell 1767 in the general direction of the filament of objects leading back to M 101. This will be an important property of galaxy clusters which appear to be involved in alignments with a central galaxy. Further evidence of this effect will be discussed later in this paper.

THE GAMMA RAY BURST

The plus sign in the SE corner of Fig 1 marks the location of a gamma ray burst which has been identified with an optical transient showing a blue continuum in which metal lines of $z = 1.477, 1.157$ and possibly 0.928 are observed (Jha et al. 2001). It is apparent that this object lies along the SE extension of the line of active objects coming from M 101, which includes Abell 1904 and 3C 295. The feature which identifies it most strongly with M 101, however, is that it was, up to June of 2001, the

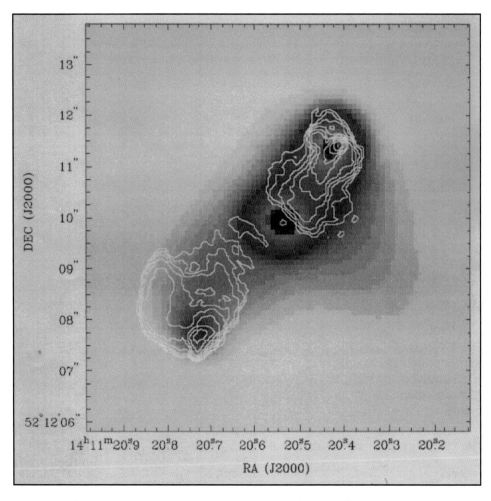

Fig. 2 - An X-ray image of 3C295 from Chandra showing X-ray hot spots being ejected at about p.a. = 144 deg. Radio contours from MERLIN are overlayed showing radio ejection in the same direction (Harris et al. 2000).

brightest GRB detected with Beppo Sax. Although a number of alignments in the *Catalogue* associate gamma ray bursters with active galaxies, it is certainly impressive that the brightest one is here associated with the nearest active galaxy ever investigated.

3C 295

To the SW of M 101 in Fig. 1 the famous radio galaxy 3C 295 is plotted. The remarkable property of this location is that it has nearly the same separation, but on the other side of M 101 from Mrk 273. Most striking, however, it is almost exactly on a line to the SE Abell cluster, at z = .071. I

150

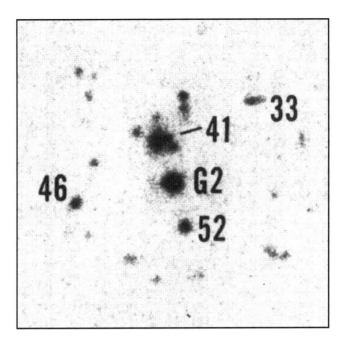

Fig. 3 - Photograph of the central galaxies in 3C295 on a IIIaJ 4m Mayall reflector plate taken by H. Spinrad. No. 41 is the radio galaxy at z = .461, No. 52 is an emission line companion at z =.467. The brighter galaxy falling exactly between them has z = .285.

remember the excitement at Mt. Wilson and Palomar observatories when the redshift of 3C 295 was first measured. It was among the strongest radio sources discovered in initial radio surveys, and Minkowski's 1960 redshift of z = 0.46 remained the highest redshift measured until after 1975. (See Sandage 1999 for a history of major events over 50 years at Palomar.)

The most recent observation of 3C 295 is in X-rays by Chandra. Fig. 2 shows a pair of X-ray condensations coming out of the nucleus at a position angle of about p.a. = 144 deg. This is the same alignment as the radio lobes shown as overlayed contours in Fig. 2. Since the ejection origin of radio lobes has long been accepted, and X-ray jets often are found at their core, the X-ray sources in 3C 295 are indicated to be in the process of ejection. It is perhaps significant that the position angle of the X-ray source ejection in 3C 295 is about p.a. = 144 deg., not far from the direction to Abell 1904 which is about p.a. = 153 deg. There is no apparent reason why they should be so aligned, but we have seen similar alignment of radio lobes from 3C 321 along the direction back to Arp 220, the third major ultraluminous galaxy (Arp 2001).

Even more curious is the fact that, at a very small scale, between 3C 295 at z = .461 and its closest major companion at z = .467, there is a bright galaxy at z = .285 (marked G2 in Fig. 3). The two higher redshift galaxies, as shown in Fig. 3, are almost perfectly aligned across the lower redshift galaxy at only 11 arcsec on either side of it. There is only a very small a priori chance of a galaxy this bright accidently occurring at this exact spot. This of course is the quintessential pattern of AGN's ejected from a lower redshift galaxy (often interpreted as gravitational lensing). There is some hint of luminous material connecting the central galaxy to 3C 295, and deeper, high resolution images should be obtained to check this possibility.

A GALAXY OF Z = 2.72

Not conspicuously associated, but in the general SW direction from M 101, is another example of the 3C 295 kind of configuration, shown in Fig. 4. There it is seen that the starbursting galaxy cB58 (MS 1512+36) has an extraordinarily large redshift for a galaxy, z = 2.72. But it is one mem-

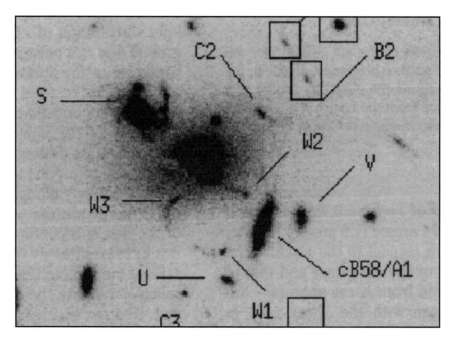

Fig. 4 - The galaxy cluster MS 1512+36 imaged in co-added V and R. (The central portion of the field is reproduced from Seitz et al. 1998 who analyzed the system as a gravitational lens.) The central, large galaxy has z = .372 and the high surface brightness starburst, galaxy cB58 has z = 2.72. The latter is 5 arcsec from the central galaxy. The irregular galaxy 3 arcsec on the other side has no reported redshift.

ber of a conspicuous pair across a central galaxy of $z = .373$. The galaxy that is the other member of the pair is thought to be a spiral (although in the Figure it seems to have more of a crab-like shape). Its spectrum was taken, but no redshift has been reported (Abraham et al. 1998). The $z = 2.72$ galaxy is only 5 arc sec from the $z = .373$ galaxy on the sky. The chance of finding such an unusual galaxy accidentally this close with this alignment is ridiculously small. It might also be remarked that just a glance at the numerous redshifts measured in this cluster will reveal that they range from $.2 \leq z \leq .6$. It would seem on the face of it to be a cluster of discordant redshift galaxies.

This system has only been interpreted in terms of gravitational lensing (Seitz et al. 1998). Aside from the criticisms implicit in the associations of supposed gravitational lenses in the present *Catalogue*, however, the lensing deformations expected in the $z = 2.72$ galaxy do not appear to match the observed form of the galaxy. The published lensing calculations are discussed elsewhere (Arp and Crane 1992).

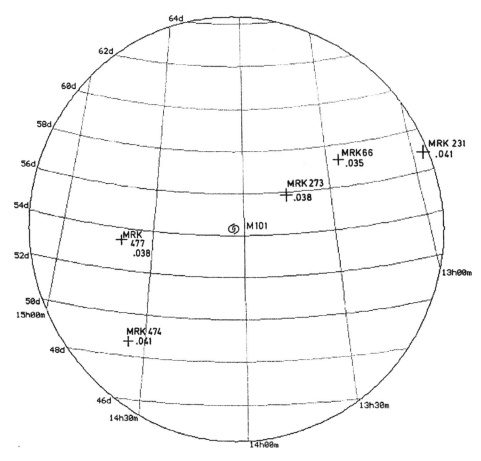

Fig. 5 - Within a radius of 10 deg., all Markarian galaxies brighter than 15.3 magnitude and z between .035 and .045 are plotted.

THE DISTRIBUTION OF BRIGHT MARKARIAN GALAXIES

Because of the provocative location of Mrk 273, it was tempting to see if there were similar objects within the extended field of M 101. A SIMBAD screen was set up to tabulate objects brighter than about 15.5 mag. with redshifts between 0.35 and 0.41. A field of radius 10 degrees was arbitrarily set. Fig. 5 shows that a total of five Markarian galaxies were found. Unexpectedly, the exceptionally active Mrk 231 showed up near the NW edge of the field. Mrk 231 is the second of what are believed to be the three most luminous infrared galaxies in the sky (Arp 2001). *Finding it here by chance, close to Mrk 273, and in the same NW direction from M 101 would appear to be quite unlikely.*

The second point Fig. 5 illustrates is that the five Markarian galaxies are distributed roughly along a NW–SE line which, as we shall continue to see,

154

is where most of the galaxies and active objects are situated. But these Markarian galaxies are all exceptional objects. The intensity with which they have been studied attests to their importance—for example the number of literature references for Mrk 231 is 509, for Mrk 273, 274; for Mrk 477, 126; for Mrk 474, 60 and even for the least studied, Mrk 66, there are 32 references.

Perhaps even more impressive, however, is that this is a general alignment, over almost 20 degrees in the sky, of exceptionally active objects with nearly the same redshift: $z = .041, .041, .038, .038, .035$. In terms of conventional redshift distance, this would represent a narrow filament of physically related galaxies spanning an enormous distance in space. The conventional view would be that these are all galaxies at one particular time in their evolution. Associated with M 101, however, they could have an origin in a single ejection event. Such events are recurrent and could later furnish different lines of objects at different discrete stages of evolution.

THE DISTRIBUTION OF BRIGHT QUASARS

In order to separate background quasars from candidates for association with M 101 another SIMBAD screen was set at the relatively bright apparent magnitude of V = 17.1 mag. The search was supplemented by visual search of the Veron and Veron Catalogue and checks with NED lists of high redshift quasars. The quasars found are plotted in Fig. 6. The first impression is that these brightest quasars are distributed along the same general line of Markarian galaxies as just discussed. On closer inspection, there are quasars near each of the plus signs that represent the Markarian galaxies from Fig. 5, suggesting that the quasars may have originated more recently from these lower redshift, active Markarian galaxies, and not necessarily as direct ejections from M 101 itself.

Some of the high redshift quasars here have such bright apparent magnitudes that, whatever the average luminosity for this redshift may be, they are certainly bunched closely within this class—if the redshift does not indicate distance (Arp 1998b). Examples are redshifts of z = 2.63 at 17.0 mag., z = 1.86 at 16.6 mag. and z = 3.19 at 15.8 mag. (We should remark that while the z = 3.19 quasar is not near a Markarian galaxy, it is quite near an infrared, IRAS, galaxy of 15.3 mag. and z = .037, and therefore similar to the Markarian galaxies plotted here.) The 15.5 mag. object is a BL Lac type quasar, OQ 530, very bright in apparent magnitude for its class, and agreeing with the close association of BL Lac's with nearby galaxies, as reported in *Seeing Red* (Arp 1997 Table 2, and 1998a).

NUMERICAL COINCIDENCES OF REDSHIFTS

A number of objects in Fig. 6 show remarkable numerical agreements in redshifts. The quasars at z = .646, .660 and .656 stand out. It should be noted that there is an additional quasar at 17.7mag., slightly below the cutoff of 17.1 mag., which has a z = .646 and falls just SW of M 101. The latter makes a close apparent pair with the z = .660 quasar directly across M 101. The wider pair at z = .646 and .656 are very closely matched in magnitude as well as redshift, but are probably associated with the two Abell Clusters that are even better centered and aligned across M 101. As remarked in the previous section, quasars at greater distances from M 101 may arise from secondary ejections from earlier ejected, still active galaxies such as Markarian objects or X-ray clusters.

The numerical coincidence of the 3C quasars at z = .961 and .967 also stands out. Undoubtedly associated with these two is 4C53.28, just SE of

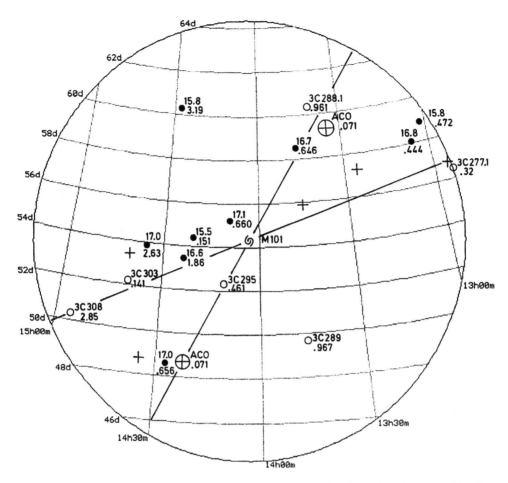

Fig. 6 - The plus signs are the same Markarian galaxies from Fig. 6, now with all QSO's less than z = 17.1 mag added as filled circles. All known 3C radio objects in the area are represented by open circles. The two circled plus signs are Abell Clusters from Fig. 1.

M 101 at mag. 17.3 to 17.5 and z = .976. In general many of the redshifts in the M 101 extended field agree very closely with peak values of redshifts in the Karlsson formula, which expresses the empirical relation found for many years for bright quasar redshifts:

$$z = .06, .30, .60, .96, 1.41, 1.96, 2.64, 3.48...$$

The agreement between these predicted values and the values for the majority of quasars in Fig. 6 is evident. Recently Burbidge and Napier (2001) have demonstrated a very significant extension of the Karlsson series to the highest redshifts in various samples of quasars.

157

THE NEAREST, MOST WIDELY SPREAD PAIR OF QUASARS IN THE SKY?

It was the non-professional astronomer Robert Harman who had the courage to look beyond the 10 degree radius I had provisionally assigned as the area of association with M 101. As Figure 7 shows, there is a pair of extremely bright apparent magnitude quasars (14.0 and 14.3 mag.), closely aligned across M 101 with very closely matching redshifts of $z = .087$ and $.089$. They are at a distance of 9.5 degrees N and 10.5 degrees S of the central M 101. It should be stressed that it is extremely unusual for such low redshift quasars to have bright enough apparent magnitudes to place them in the conventional quasar class luminosity. (If they were at usually supposed redshift distances they would have absolute magnitudes of $M = -24.7$ and -24.1, where the arbitrary class limit is -23.) Adopting the usual criteria of closeness for their apparent magnitude, centering, alignment, similarity of quasar redshifts and apparent magnitudes would produce an estimated probability of chance association $\leq 10^{-7}$. Together with the many previous cases this would take the overall probability for physical association of quasars with low redshift galaxies beyond sensible calculation.

THE 3C RADIO SOURCES

Again in Fig. 6 we see that the brightest, earliest discovered, 3C radio sources fall mostly along the line of objects which passes through M 101 from NW to SE. Apparently 3C277.1 is associated with Mrk 231, and 3C 295 apparently with M 101 itself. Some of the others may also be associated with nearby active galaxies such as Mrk 477.

3C 303 is a particularly well studied object because it has a prominent one sided radio jet pointing from a compact radio galaxy of $z = .141$ toward its double radio lobe, the northern component of which is at p.a. $= 280$ deg. (Lonsdale et al. 1983). The direction to M 101 is about 289 deg. The ROSAT PSPC (low resolution X-ray photon counter) observation shows X-ray sources mildy extended in the NW-SE direction. Of particular importance, however, are three ultraviolet excess objects apparently associated with the western lobes of 3C 303 (Kronberg et al. 1977). From the spectrum of one Margaret Burbidge obtained a redshift of $z = 1.57$. In spite of the fact that this association was discussed as possible conclusive proof that quasars are not at great distances, no further time has been assigned to obtain the spectra of the remaining two candidates.

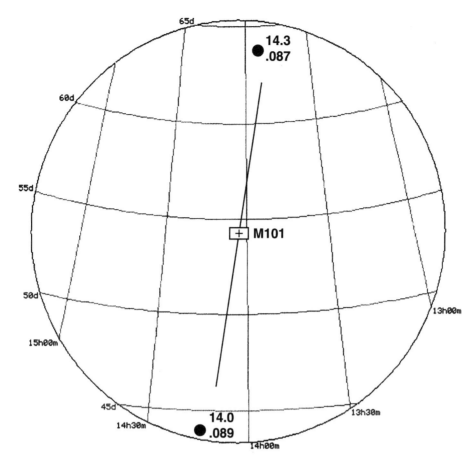

Fig. 7 - Within a radius of 11 degrees bright apparent magnitude quasars are plotted. The quasars PG 1411+442 and PG 1351+64 are at 10.5 and 9.5 deg. radial separation and aligned at p.a. = 169.5 and 173.3 on either side of M101.

With the plot in Fig. 6 now complete, we could now propose that there was a cone of ejected objects from M 101 in the NW-SE direction. Or we could interpret the distribution of objects as two rather narrow lines, one at p. a. = 110 and the other at p.a. = 147 deg. We will prefer the latter interpretation, because we now wish to examine the possible connection of the Abell galaxy clusters with these lines of objects.

159

From previous evidence we have some expectation of galaxy clusters being elongated along the line of their apparent ejection. The clusters Abell 3667 and 3651 have been found to be paired across the central galaxy, ESO 185-54 (Arp and Russell 2001). The clusters were strongly elongated along this connecting line, and later Chandra observations actually showed a bow shock (later described as a "cold front") moving outward along just this line in Abell 3667 at 1400 km/sec. (See also Figs. 5 and 6 in the following Appendix B.) The famous "gravitational arc" cluster, Abell 370 showed elongation back toward its purported galaxy of origin, the bright Seyfert NGC 1068. Several other similar examples were also noted in the above reference.

Catalogued galaxies within a radius of 60 arcmin around the cluster Abell 1904 are shown in Fig. 8. The cluster is noticeably elongated in a direction leading back to 3C 295, and hence very closely along the line back to M 101. Inspection of PSPC X-ray measures reveals a flattened core to the cluster with faint material trailing out perpendicular to this core in a direction back toward M 101. (See Fig. 13, in Appendix B which suggests an ablating passage through an intergalactic medium.) Since that PSPC exposure was only 3.8 kilosec, it would be important and possibly decisive in determining ejection origin from M 101 if a higher resolution Chandra observation of this cluster could be obtained.

Another cluster of somewhat fainter galaxies, Abell 1738, with z = .115, is not plotted in Fig. 5 or 6, but lies just N of Mrk 66, and is shown close-up in Fig. 9. This cluster has relatively few cluster members with measured redshifts, but the figure shows that they are aligned back toward M 101. It is located on the edge of a PSPC exposure which can only establish that it is also an X-ray cluster. Not much can be said about Abell 1767 except that it gives some impression of being aligned back toward M 101 at about 140 deg. The PSPSC X-ray sources around 3C 303 appear inclined at about p.a. = 110 deg. toward M 101. However, quantitative determination of optical galaxy isopleths for these last two clusters and/or X-ray contours of their smoothed distribution would be required.

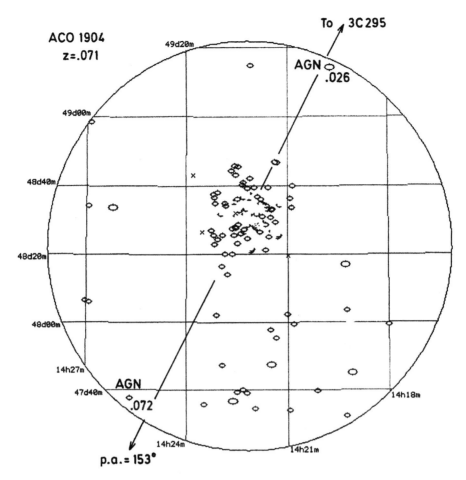

Fig. 8 - All catalogued galaxies from SIMBAD within a one deg. radius centered on Abell cluster 1904. The line at p. a. = 153 deg. points back to 3C295, and from there to M101. Many of the galaxies in the SW quadrant have redshifts near z = .07, and there are also some AGN's and quasars. (See also Fig. 13 in Appendix B.)

The shapes of the clusters shown In Figs. 8 and 9, however, in conjunction with similar data in the *Catalogue* and Appendix B, would seem to furnish compelling evidence that the cluster originated from a point near the present M 101. In the conventional view, constraints that give spontaneously forming, linear, non-equilibrium configurations in deep space would seem to difficult enough. But when these point back to a large galaxy, a connection with generally linear, recurrent ejections from that galaxy seems unavoidable.

161

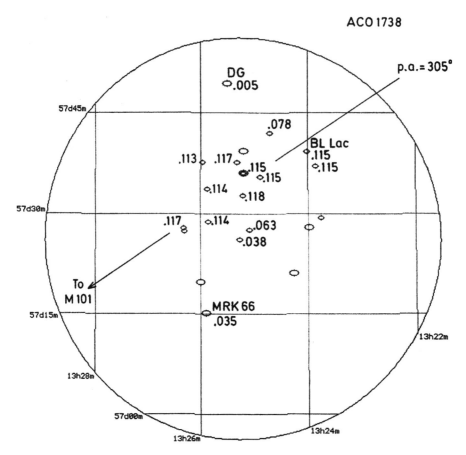

Fig. 9 - All catalogued galaxies from SIMBAD within a 30 arc min circle centered between Mrk 66 and Abell 1738. The line at p. a. = 125 deg. points back to M101.

SCALE OF THE M 101 ASSOCIATIONS

The large scale of the associations found around M 101 raises the question whether they are compatible with the many similar associations found around more distant active galaxies. The distance to M 101 is, from Cepheids, m–M = 29.13 mag., or 6.7 Mpc. (Freedman et al. 2001). If the majority of the previous associations were made with central galaxies at about the distance of the Local Supercluster, i.e., 15-16 Mpc, then they would extend out to a factor of 2.3 greater radius on the sky at the closer distance of M 101. Since the observed extensions of galaxy clusters and low redshift AGN's around central galaxies were in the range of 2-4 degrees (Arp and Russell 2001), the range around M 101 would be expected to be from about 5 to 10 degrees. This is roughly the angular ex-

162

tent of the associations with M 101 which were found in Figs. 1, 4 and 5. As an aside, it is a general property of the associations that the higher redshift quasars are found closer to their galaxy of origin and evolve to lower redshift as they move out to become companions at greater radii. There can, however, be more recent, secondary ejections from these outlying active companions.

It is important to note that the apparent magnitudes of the low redshift AGN's around the more distant associations tended to be in the 17 mag range, the quasars in the 19 mag. range and the very high redshift quasars in the 20th mag. and fainter range. At the closer distance of M 101 they would be expected to be about 2 mags. Brighter; as can be seen from the figures and text, that is just about what is observed around M 101. There are arguments, however, that M 101 belongs to the M 81 group at about 3.6 Mpc distance (Arp 2002). In that case the scale factor would be still larger.

If M 101 is considered as the central galaxy the strongest evidence for activity in the galaxy is the long straight arm in the outer NE quadrant. It was classified as a spiral galaxy with one heavy arm—a fact which led to its identification as No. 26 in the *Atlas of Peculiar Galaxies* (Arp 1966). At that time I argued that spiral arms generally represented ejection tracks, and that ScI galaxies were a younger and particulary active kind of galaxy. The nucleus of M 101 is not now particularly active spectroscopically, but then such activity is probably intermittent. The far-flung streams of older galaxies and low redshift AGN's associated with M 101 would suggest that they originated from activity in the rather distant past.

Sunyaev-Zeldovich Effect

Calculation of the distances of galaxy clusters from their scattering of microwave background in conjunction with measurements of their X-ray surface brightness seems to rest on such proven physical principles that it is difficult to see how anyone could accept much closer distances, as the observations in the present paper claim. If we are, however, to give the observations even equal consideration we must face the apparent discordance of nearby galaxy clusters with S-Z distance determinations.

Perhaps the first point to be made is that galaxy clusters which are nearby would present strongly non-equilibrium physical conditions. If they were formed in more recent ejections from active galaxies, and they themselves are ejecting secondarily and in non equilibrium configurations, perhaps it is incorrect to assume equilibrium temperature, radiation and energy densities. A strong indication that non-equilibrium conditions apply is provided by cooling flows. In many clusters the densities near the center are so high that the cooling flows would exhaust the available heat on much less than a cosmic time scale. The problem is so acute that suggestions have been made for merging or accretion of companion objects to resupply the energy in the center (A. Fabian, Moriond Conference).

Actually the situation in many clusters, especially those dominated by large central galaxies, is that powerful ejections take place that must intermittently add bursts of energy into the surrounding cluster medium. (See e.g., 3C 295 pictured here in Fig. 2.) The question then becomes: Do observations of the cluster medium at any given time represent an equilibrium physical state on which the conventional physical calculations can be made?

To make this suggestion more specific, consider that in order to account for the intrinsic redshifts of the objects of various ages associated with an active galaxy, it has been necessary to make a more general solution of the general relativistic field equations where particle masses are a function of time (Narlikar and Arp 1993). This proposal requires initial ejection a plasma of low mass particles near light velocity. Because of their low mass, the newly ejected particles can have large scattering cross sections, enabling microwave photons to be efficiently boosted. The synchrotron/*bremsstrahlung* jets, however, are observationally well collimated which must mean relatively low temperatures orthogonal to the direction of travel. Their energy is mostly converted to temperature only at the interaction surface of the jet cocoon with the surrounding medium. Both of these—the strong scattering and low temperature effects—would tend to give a large distance for a small total mass cluster in the S-Z equation.

This proposal can be put in a possibly observable way: even if there were a nearby cluster in temporary temperature equilibrium with a well-determined X-ray surface brightness, but no measurable S-Z microwave depression, could an upper limit for the distance of this cluster be calculated?

Conclusion

By now it has been long accepted that radio jets and lobes, often accompanied by X-ray jets and X-ray emitting material, are ejected, in generally opposite directions, from active galaxies. Since compact, energetic X-ray sources must be relatively short lived, it is natural to suppose that most of the excess density of X-ray sources around active galaxies (Radecke 1997) has been ejected recently. Their identification with more highly redshifted quasars, and subsequent evolution into lower redshift quasars, compact active galaxies and normal galaxies has been discussed elsewhere (Arp 1998a). The recently discovered ejection of higher redshift quasars and X-ray galaxies from Arp 220 further supports this conclusion. (See Fig. 10 in Introduction and Arp 2001; Arp et al. 2001.)

In the vicinity of M 101, examined in detail in this paper, we have the rare opportunity to see associations closer than in most nearby galaxies by a factor of more than 2. Using the principle that the brightest companion objects will cluster around the brightest parent galaxies, we have identified lines and filaments of higher redshift and active objects extending out of, or through, M 101. In the course of this survey we have had the pleasure of encountering many of the most renowned and intensively observed Markarian and 3C radio objects, and some very bright apparent magnitude quasars. It is gratifying to rediscover the long strings of low and intermediate redshift galaxies I first noted in 1984 (Arp 1984) and revisited in 1990, shown here in Fig. 1. These conspicuously aligned strings of galaxies remind one of the features presently explained with various cold dark matter scenarios. Here, however, their explanation would rely on evolving high-energy ejections from a variety of central galaxies. The marked elongation of clusters would then suggest formation by ablation from ejected plasma knots. The famous M 87 jet visually gives this strong impression.

We have also seen some striking examples of quantized redshifts around the low redshift central galaxy M 101. M 101 has such a low redshift that no corrections are needed. But as noted previously in the *Catalogue*, when the parent galaxy has an appreciable redshift, the redshift of the ejected object must be corrected to its rest frame. The conventional redshift-distance interpretation of the quantized redshifts would, of course lead to a universe of concentric shells centered on our galaxy. The alternative interpretation of the redshifts as a measure of evolutionary stage, however, would permit the quasars to be born in separate ejections, which may take place in periodic events.

REFERENCES

Abraham, R., Yee, H., Ellingson, E., Carlberg, R., Gravel, P. 1998, *ApJS* 116, 231

Arp, H. 1966, *Atlas of Peculiar Galaxies*, California Institute of Technology; *ApJS* 123, Vol. XIV

Arp H. 1984, P.A.S.P. 96, 148

Arp H. 1990, *J. Astrophys. Astr.* (India) 11, 411

Arp, H. 1997, *A&A* 319, 33

Arp, H. 1998a, *Seeing Red: Redshifts, Cosmology and Academic Science*, Apeiron, Montreal

Arp. H, 1998b, *ApJ* 496, 661

Arp, H. 1999, *ApJ* 525, 594

Arp, H. 2001, *ApJ* 549, 780

Arp, H. 2002, *ApJ* 571, 615

Arp. H, and Crane, P. 1992, *Physics Letters* A 168, 6

Arp, H. and Russell, D. 2001, *ApJ* 549, 802

Arp, H., Burbidge, E., Chu, Y., Zhu, X. 2001, *ApJL* 553

Brandt, W., Hornschmeier A., Schneider, D., et al. 2000, AJ 119, 2349

Burbidge, E. M. 1995, *A&A*, 298, L1

Burbidge, E. M. 1997, *ApJ*, 477, L13

Burbidge, E. M. 1999, *ApJ*, 511, L9

Burbidge, E. M. 1995, *A&A*, 298, L1

Burbidge, E. M. 1997, *ApJ*, 477, L13

Burbidge, E. M. 1999, *ApJ*, 511, L9

Burbidge, G. and Napier W. 2001, *AJ*, 121,21

Chu, Y., Wei, J., Hu, J., Zhu, X. and Arp, H.C. 1998, *ApJ* 500, 596

Harris, D. et al. 2000, *ApJ* 530, L81

Jha, S., Pahre, M., Garnavich, P. et al. 2001, *ApJ* 554, L155

Kronberg, P., Burbidge, E., Smith, H., Strom, R. 1977, *ApJ* 218, 8

Hickson, P. 1994, *Atlas of Compact Groups of Galaxies*, Gordon and Breach

Lonsdale, C., Hartley-Davies R., Morison, I. 1983, *M.N.R.A.S.* 202, 1L

Narlikar J., Arp H. 1993, *ApJ* 405, 51

Narlikar, J. and Arp, H. 2001, *ApJ*, to be submitted

Radecke, H.-D. 1997, *A&A* 319, 18

Sandage, A. 1999, *ARA&A* 37, 445

Seitz, S., Saglia, R., Bender, R., Hopp, U., Belloni, P., Ziegler, B. 1998, *M.N.R.A.S.* 298, 945.

Appendix B

FILAMENTS, CLUSTERS OF GALAXIES AND THE NATURE OF EJECTIONS FROM GALAXIES

INTRODUCTION

Classic cases of ejected radio sources were recognized early by the lobes extending in opposite directions from Virgo A and Cen A (M 87 and NGC 5128). A good example is Plate 8-18 in *Seeing Red* (Arp 1998a). When X-ray observations became available, however, it was seen that narrower cores of X-ray jets extended inward from these radio jets. Lower power radio jets are now showing similar X-ray cores (e.g., Pietsch and Arp 2001), particularly with the higher resolution Chandra observations (Wilson et al. 2001; Worral et al. 2001). One obvious implication is that the higher energy density X-ray jets supply synchrotron particles, which become the outer, lower frequency envelope of the radio lobes as they age.

It was also evident that X-ray jets tended to break up into discrete knots as they emerged from the nucleus of the ejecting galaxy. The knots resembled the compact X-ray sources that were found farther out, characteristically paired across galaxies with active nuclei (Arp 1998a). These X-ray sources, when measured spectroscopically, were invariably identified with medium high redshift quasars. In fact, the number of pairings of radio quasars was soon outstripped by the more frequent cases of pairing of X-ray quasars across active galaxies. This result brings us to another empirical suggestion, namely, that the radio material, being observably more diffuse than the optical and X-ray material, is stripped away from the more compact X-ray condensations as it passes through the galactic and intergalactic medium. Examples of this are well illustrated by 3C 212 (Stockton and Ridgeway 1998) and Mrk 231 (Arp 2001). The clouds of radio material are then much less accurate markers of the presence of energetic quasar cores, are generally widely spread around active galaxies, and have less frequent optical identifications.

The key to the origin and nature of ejected material must, therefore, lie in the X-ray jets. The jets are characteristically featureless in optical spectra, but after forming stars, the optical objects observed in the outer extensions of the X-ray ejections invariably show excited atomic spectra of variously higher redshifts. We discuss here examples of lines of discrete X-ray sources leading from the centers of large, active, low redshift galaxies. If the physical reality of the lines is accepted, then they suggest an evolutionary sequence originating in the creation of young matter in the nuclei of large active galaxies and proceeding through maturation into normal, neighboring companion galaxies.

The final sections of this Appendix give a more detailed examination of some particularly striking examples of elongated clusters emerging from low redshift central galaxies. We then conclude with a physical interpretation of the nature of the ejected X-ray plasma which might explain both the observed behavior of the ejected material and the physical cause of the time varying, intrinsic redshift.

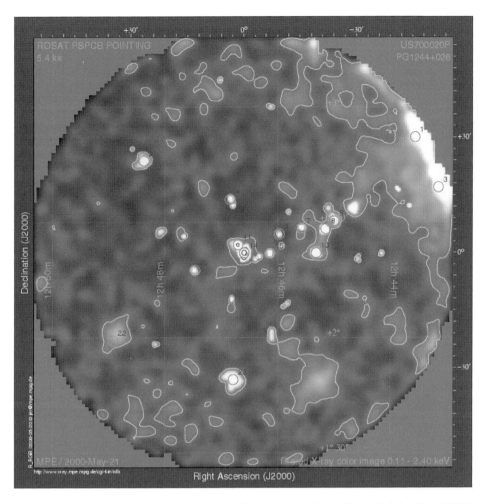

Fig. 1a (Plate 1a). Full, approximately 50 arcmin radius, X-ray field of the PSPC detector on ROSAT. A line of bright, compact X-ray sources connects NGC 4636 (z = .003) at upper right with the object in the center, the AGN PG1244+026 (z = .048). (See Fig. 1b.)

X-RAY SOURCES CONNECTING PG 1244+026 AND NGC 4636

Fig. 1a shows a 5,486 sec ROSAT, PSPC exposure of the AGN/Seyfert, PG 1244+026 (S1n). The color picture is from the standard (ROSAT source browser) reduction. It can be seen that about a dozen sources form a line leading from the low redshift, active galaxy NGC 4636 to PG 1244+026. This is the primary observational evidence, and if it is judged to be significant, we have a physical connection between an active galaxy at

171

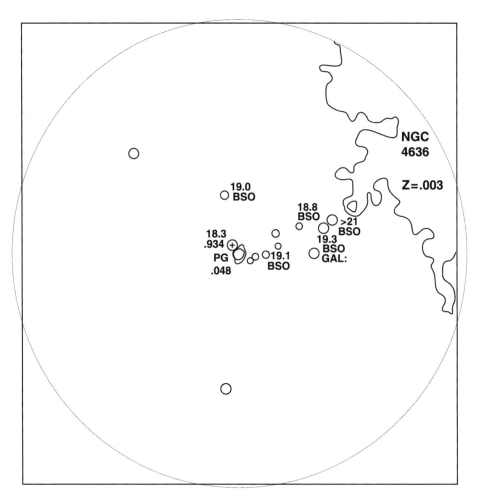

Fig. 1b. A schematic representation of the bright sources in Fig. 1a. Those sources that have been identified with optical objects are almost all quasar candidates (BSO's or blue stellar objects). Their V mag.'s are labelled and their positions are given in Table 1. Redshifts of known quasars are also labelled.

cz = 979 km/sec and a quasar-like object of cz = 14,400 km/sec (z = .003 and .048).

Recent Chandra X-ray observations of NGC 4636 show a spectacular example of X-ray emitting ejected arms. The connection with the line of quasars and quasar candidates shown here enables us to study the nature of this ejection. To this end the brightest X-ray sources in the line have been examined on deep Schmidt survey plates. As the schematic representation in Fig. 1b shows, 6 of the brightest sources have been optically identified. Most turn out to be blue stellar objects (BSO's) of the kind which can later be confirmed to be quasars or AGN's. This expectation is reinforced by the fact that one of them, PG 1244+026, has already been measured as an AGN of $z = .048$, and another as a quasar of $z = .934$.

TABLE 1 OBJECTS IN FIG 1B.

No	R.A. 2000	Dec.	Rmag.	Bmag.
6	12 44 59.8	02 30 23	--	21
7	12 45 33.8	02 28 25	18.8	18.9
8	12 45 07.2	02 28 18	19.3	20.0
15	12 46 07.4	02 21 53	19.1	18.8
13	12 45 16.8	02 21 59	18.5	18.1*
4	12 46 50.9	02 36 04	19.0	19.3

*BSO/gal

Of course the redshifts of the remaining candidates should be measured because they will give further data on the physical composition of the emerging filament. But some preliminary comments about the nature of the ejection from the primary galaxy may be permitted here. First, however, it will be useful to buttress the reality of the physical connection by presenting some observations of other, independent but similar configurations.

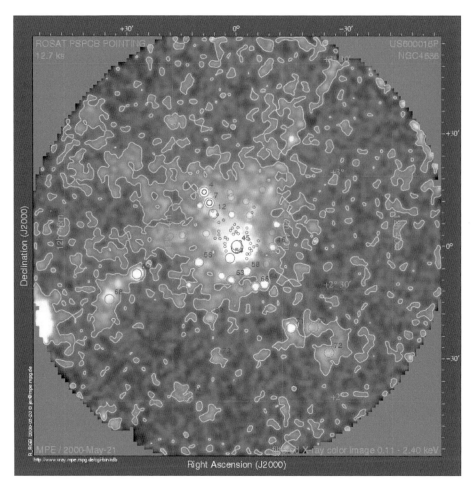

Fig. 2a (Plate 2a). Full, approximately 50 arcmin radius, X-ray field of the PSPC detector on ROSAT. A line of bright compact X-ray sources extends in either direction from NGC 4636 (a bright X-ray, E galaxy with z = .003). (See Fig. 2b.)

ADDITIONAL X-RAY EJECTIONS FROM NGC 4636

NGC 4636 is interesting because it is such a strong X-ray source, and a long PSPC exposure of 13,070 sec is available. Fig. 2a shows the standard reduction of this exposure. It is now apparent that there is a line of about 7 or 8 strong X-ray sources oriented NNE to SSW across the nucleus. Again, these have been examined; they are listed in Table 2 and tentative identifications labeled in Fig. 2b.

Several comments can be made at this time. First, recent Chandra observations (Jones et al. 2002) show strong X-ray ejections from the central

174

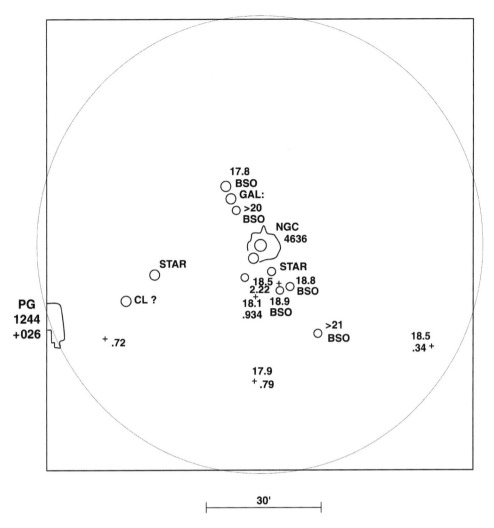

Fig. 2b. A schematic representation of the bright sources in Fig. 2a. Those sources that have been identified with optical objects are almost all quasar candidates (BSO's or blue stellar objects). Their V mag.'s are labelled and their positions are given in Table 2.

arcmins of NGC 4636 in a roughly NE-SW direction. But earlier, Matsushita et al. (1998) had shown high energy ASCA X-rays extending just out to the distance and along the line of the discrete sources shown here in Fig. 2b. It appears to be a classic ejection in opposite directions across an active nucleus. Now, however, in the outer reaches, this ejection is seen to contain individual X-ray sources which are identified as probable quasars.

Secondly, NGC 4636 is a bright (B_T = 10.50 mag.), nearby galaxy, and its associated quasars and candidates are rather bright in apparent magni-

175

TABLE 2 OBJECTS IN FIG 2B.

No	R.A. 2000	Dec.	Rmag.	Bmag.
4	12 43 25.9	02 55 49	17.8	17.0
7	12 43 20.0	02 52 56	15.4	16.7[*]
12	12 43 15.8	02 49 59	---	20.0
64	12 42 19.1	02 31 19	18.8	18.5
66	12 42 28.4	02 30 44	18.9	18.4
71	12 41 50.2	02 19 31	---	21

[*]Galaxy

tude—in the 17.8 to 18.9 mag. range. In Fig. 2b additional non X-ray se-lected quasars from the LBQS survey in the same apparent magnitude range can be noted. This is more or less the same range as in the line leading to PG 1244+026. Very intriguing is the presence of a quasar just S of NGC 4636 at $z = .934$, and another just next to PG 1244+026 at ex-actly the same $z = .934$, both having nearly the same apparent magni-tude. In the same field it is not unusual to detect some quasars as X-ray sources, as well as some from objective prism and other kinds of surveys, although it is interesting to note that the candidates strong in X-rays are closer and better aligned along the X-ray ejection.

Two of the candidate quasars in the NGC 4636 line and one in the PG 1244+026 line are optically faint—in the $V \geq 20$ and 21 mag. range. There is precedent for this, for example in the BL Lac object just N of, and connected to, the nearby Seyfert NGC 4151 (Arp 1997). It is very bright in X-rays and faint optically, suggesting that this represents a temporary phase in its evolution, perhaps related to the strong optical variability of some quasars.

Overall, however, the disposition of strong X-ray sources around NGC 4636 indicates that this nearby, massive and notably active galaxy is ejecting AGN's of higher redshift. A prominent line is shown in Fig. 2a and b, and another line leading to PG 1244+026 is shown in Fig. 1a and b. A number of cases of nearly perpendicular lines of X-ray sources ema-nating from active galaxies have been shown in an earlier publication (Arp 1996: see particularly Figs. 14 and 15 in that reference).

LINE OF X-RAY SOURCES FROM THE STARBURST GALAXY NGC7541

NGC 7541 is a B = 12.45 mag. starburst galaxy of redshift z = .009 with a blue jet/arm emerging from its disk. It is centered between two very bright radio sources (Arp 1968). Fig. 3 in the Introduction to the *Catalogue* shows that the two 3C radio sources were eventually measured at z = .22 and .29. A similar pair of 3C quasars across *Atlas of Peculiar Galaxies* No. 227 (NGC 470/474) with redshifts z = .672 and .765 (shown in Fig.2 of the Introduction) was calculated to have only a 2×10^{-9} probability of chance occurrence (Arp and Russell 2001). The physical association of the two radio quasars with NGC 7541 is about the same order of significance. But a map of the larger-radius field that follows NGC 7541 in the *Catalogue* (R.A. = 23^h15^m) shows that, at approximately a right angle to this radio pair, is an X-ray source with a redshift of z = .30 that is intermingled with some galaxies of about z = .13.

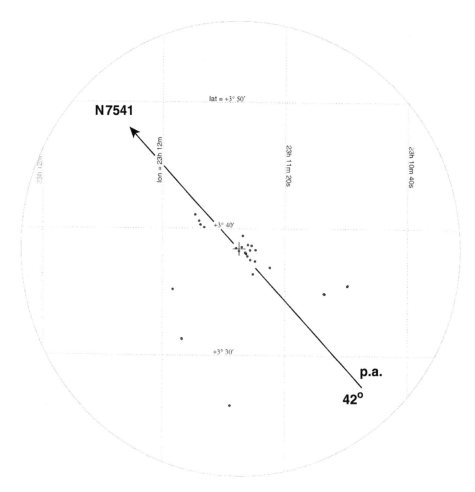

Fig. 3. A ROSAT, high resolution HRI X-ray map centered on Abell 2552. The compact sources are aligned as exactly as can be judged back toward NGC 7541. (See Figure for NGC 7541 in the main *Catalogue* and Fig. 3 in the Introduction).

There is a dominant X-ray galaxy at the center of this cluster, Abell 2552, that causes the cluster to be classified as a strong X-ray cluster at $z = .300$ (Böhringer et al. 2000; Ebeling et al. 2000). But regardless of the details of this grouping of objects, Fig. 3 in this Appendix shows that the X-ray sources in the cluster are strung out in a well-defined line. That line, as illustrated in the present figure, points accurately back to the active galaxy NGC 7541. The individual X-ray sources are relatively close together and faint, so, unlike the previous case of NGC 4636 and PG 1244+026, it is difficult to make probable optical identifications. *But it is clear that the X-ray emission, whatever its exact form, is narrowly spread along the line back to NGC 7541.*

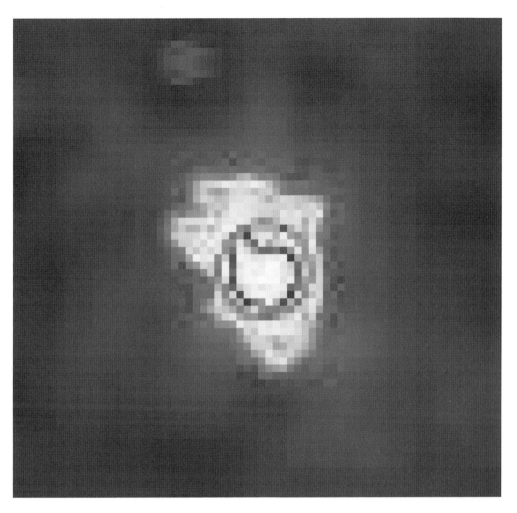

Fig. 4 (Plate 4). This is an enlarged view from the high resolution, HRI X-ray image. X-ray material appears to emerge from this central galaxy and extend out along the p.a. = 42 deg. line back toward NGC 7541.

Further, it is very important to note that the high resolution HRI map of the central AGN also shows extended emission along p.a. 42 deg. The material extending from the active, dominant galaxy is shown well in the enlargement in Fig. 4 (Plate 4). It is becoming clear that one extreme in a wide range of cluster morphology is exemplified by a very dominant AGN surrounded (usually closely) by a large number of much fainter cluster galaxies. 3C 295, discussed in Appendix A, is perhaps a typical example of this kind of cluster. Abell 2552 appears to be one of this kind, a dominant radio/X-ray galaxy which is entraining (or ejecting) pieces of itself to form a cluster of galaxies around it which are much fainter but of roughly the same redshift.

179

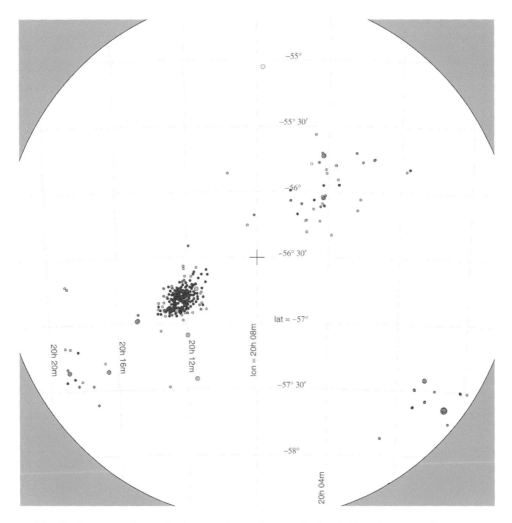

Fig. 5. The very elongated, very strong X-ray cluster A3667 is shown from PSPC exposures reduced in the 2RXP mode. See following Figure caption for further discussion (Arp 2001b).

ELONGATED X-RAY CLUSTERS

THE ELONGATED X-RAY CLUSTER A 3667

We can turn the question around and ask: If we look in the vicinity of strongly elongated X-ray clusters do we find them pointing to lower red-shift, active galaxies? A systematic survey should be conducted, but some spectacular examples can already be mentioned. An almost line-arly elongated cluster pouring out 2440 X-ray counts/ksec is shown in

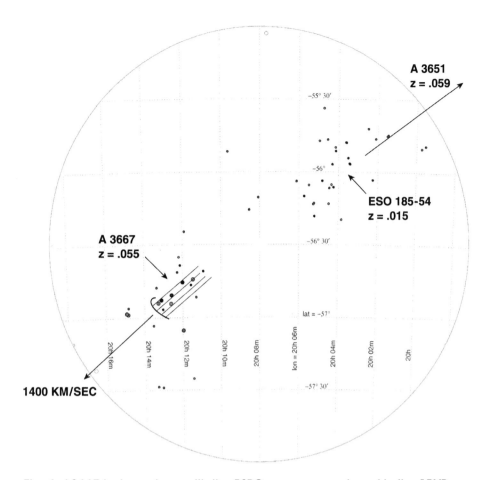

A 3651
z = .059

−55° 30′

ESO 185-54
z = .015

A 3667
z = .055

−56°

−56° 30′

lat = −57°

lon = 20h 06m

20h 16m

20h 14m

20h 12m

20h 10m

20h 08m

20h 04m

20h 02m

20h

−57° 30′

1400 KM/SEC

Fig. 6. A3667 is shown here with the PSPC exposures reduced in the 1RXP mode in order to emphasize the brightest sources in the cluster. The X-ray E galaxy ESO 185-G4 (m_B = 11.19 mag.) is at z = .015 and extends in either direction towards the A3667 and A3651 clusters at z = .055 and .059. (See Arp 2001b for further details.) Recent Chandra observations show a cold front moving down the length of the cluster, away from ESO 185-G4, with a velocity of 1400 km/sec.

Fig. 5. This cluster, A 3667, is paired with another elongated X-ray cluster, A 3651, across an active, m_B = 11.9 mag. E galaxy. The details of how the X-ray emission from the central galaxy and the intermingling of the cz = 4,500 km/sec galaxies with the 17,000 km/sec galaxies define the line of pairing and elongation are discussed elsewhere (Arp and Russell 2001).

In Fig. 6 here, we emphasize a subsequent result which would seem conclusive in showing that the cluster A 3667 has been ejected from the central ESO 185-G4. The Chandra measures by Vikhlinin et al. (2001) show

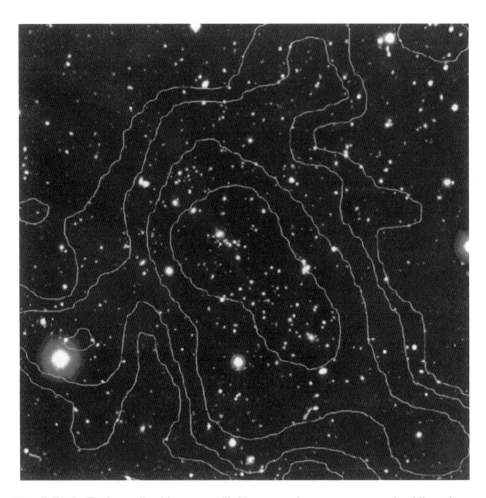

Fig. 7 (Plate 7). An optical image, with X-ray contours superposed, of the cluster RXJ0152.7-1357 at $z = .83$ (Rosati et al.). At its redshift distance it would be one of the most luminous clusters known. It points, however, nearly at the strong X-ray E galaxy NGC 720. (See next Figure.)

that a cold front is moving directly along the line of the extension of A 3667—away from the $z = .015$ group with $z = 1400$ km/sec. This information will be useful when we try to construct a model of how the initial ejection evolves into the objects we observe.

THE ELONGATED X-RAY CLUSTER RXJ 0152.7-1357

Fig. 7 (Plate 7) shows another spectacular example of a very elongated cluster. Could there be clearer evidence for non-equilibrium mass and radiation distribution? If it were at its redshift distance it would be "one of

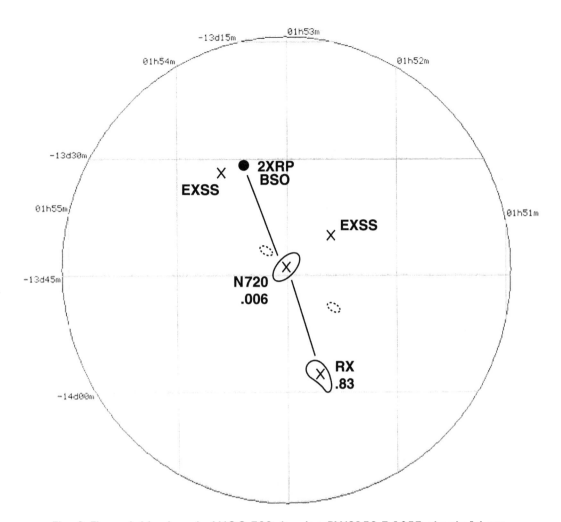

Fig. 8. The neighborhood of NGC 720 showing RXJ0152.7-1357 at only 14 arc min separation. EXSS sources (extended X-ray sources) which are likely to be clusters are shown as X's. Dashed lines indicate possible optical clusters along the minor axis of the central galaxy. NGC 720 is reported to have "twisted" X-ray isophotes emerging from its center.

the most X-ray luminous clusters known" (Rosati et al. 2000). It points within 15 deg. of direction to NGC 720, a bright $B_T = 11.15$ mag. galaxy which is a bright source in the ROSAT all sky survey (148 counts/ks) and has been studied for its "twisted" central X-ray isophotes (Buote and Canizares 1996).

As noted before (Arp and Russell 2001), the chance of finding a galaxy this bright so close to an arbitrary point in the sky in the range $20h \leq R.A. \leq 4h$ is less than 10^{-3}. But then we must decrease this further by the chance of the galaxy being such a strong X-ray source and the cluster's near alignment. The redshift of this very X-ray luminous cluster is z = .83. Its unusually small (14 arcmin) separation from NGC 720 is in accord with the tendency for the higher redshift objects to be closer to their galaxy of origin. This new case agrees very well with the strong X-ray cluster Abell 370 discussed in the earlier paper (Arp and Russell 2001). There A 370 was conspicuously elongated back toward the very active Seyfert NGC 1068, 1.7 deg. distant. The greater separation would accord with its lower, z = .375 redshift.

Fig. 8 (repeated from the main *Catalogue*) shows that there are additional EXSS (extended X-ray sources) within about 15 arcmin of NGC 720. These are likely to be X-ray clusters and should be investigated further. The dashed ovals indicate candidate optical clusters just at the limit of the POSS II plates. Deeper optical images would be capable of confirming their apparent orientation along the minor axis of NGC 720. Finally there is a strong X-ray source *exactly aligned and equidistant* on the other side of the nucleus of NGC 720 from RXJ 0152.7-1357. It is identified below with its PSPC number; it would be interesting to know its redshift:

No. 11 53^{M}23.62^s−13^{D}30^{M}59.2^s 19.0 19.0 (R,B MAG.)

A compilation of ultra luminous X-ray sources (ULX's) from HRI observations close to bright galaxies (Colbert and Ptak 2002) shows two roughly on either side of NGC 720. They would be good examples of the suggested identification of ULX's with recently ejected quasars (G. Burbidge, E. M. Burbidge and H. Arp, *A & A* 400, L17). The Colbert and Ptak finding charts for 87 such ULX's, when optical candidates are measured, should settle the association of quasars with active, low redshift galaxies. This, of course, is the same association demonstrated by Radecke (1997) and Arp (1997) for PSPC X-ray quasars around Seyfert galaxies.

184

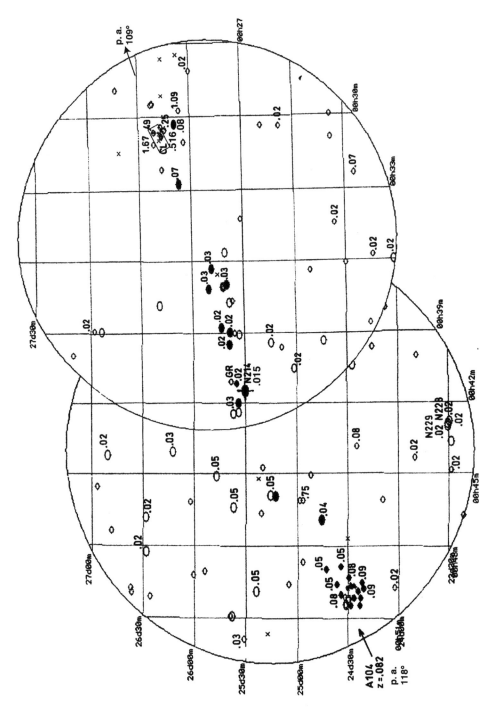

Fig. 9. This is the same map of galaxies in the NGC 214 region as presented in the main *Catalogue*, but now the 8 x 8 arc min region observed by Chandra is outlined. This area is shown in detail in Fig. 11. The central galaxy, NGC 214, is noted at z = .015 and the galaxies with .015 ≤ z ≤ .11 are filled in as solid symbols along the lines.

Additional Examples of Elongated Clusters Aligned with Galaxies of Lower Redshift

NGC 214 and RXJ 0030+2618

Fig. 9 shows a serendipitously discovered association that illustrates very clearly the elongation of a cluster back toward a central galaxy over an extended region of the sky. I discovered it when I investigated the unusual galaxy cluster CRSS J0030.5+2618. In that cluster the dominant, X-ray galaxy has $z = .516$ and a significant excess of Chandra X-ray sources (Brandt et al. 2000; Cappi et al. 2001). As discussed in the main *Catalogue*, it also has two galaxies of redshift $z = .247$, one of $z = .269$, and three quasars of $z = .492$, 1.665 and 1.372, all within 5 arcmin, and a further quasar $z = 1.094$ within 8 arc min.

I asked the question: "Where is the bright galaxy from which all this was ejected?" I found it about 2.5 degrees away. It was NGC 214, an Scl galaxy of $B_o, i_T = 12.48$ mag. and a redshift of $v_o = 4757$ km/sec or $z = .016$. This redshift indicates it is associated with the ubiquitous Perseus-Pisces filament (Arp 1990). But the striking feature was that bright companion galaxies at $z = .02$ and .03 stretched away to the WNW, directly toward the cluster of high redshift objects centered around the CRSS cluster.

Naturally I then asked: "What is on the other side of NGC 214?" As Fig. 9 shows there is an Abell Cluster, ACO 104 about 2 degrees away and nearly opposite NGC 214 with a redshift of $z = .082$. The striking feature of this cluster is that it is clearly elongated back toward NGC 214.

ABELL 104

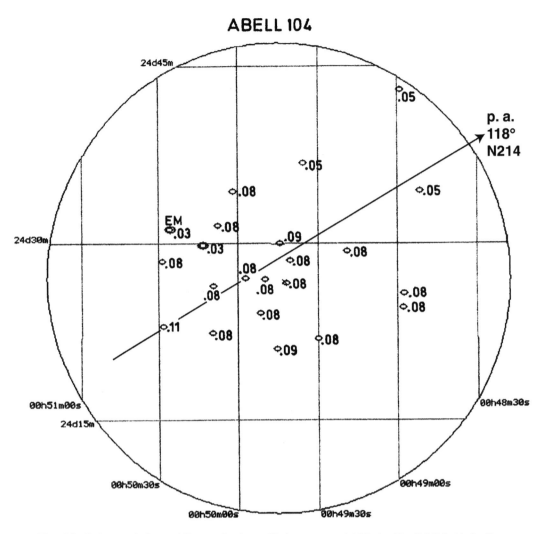

Fig. 10. Enlarged view of the galaxies with known redshifts in Abell 104. Note the galaxies at z = .03, one an emission line object. Redshifts similar to that of the central object, NGC 214, seem to be mixed into the area of this cluster. The line back to NGC 214 runs at p.a. = 298 deg.

Also interesting is the fact that galaxies of z = .04 and z = .05 lead into the cluster from the direction of NGC 214. (There are even two galaxies of z = .03 inside the cluster as shown in Fig. 10.)

In all these cases all the galaxies and optical objects in these areas should be completely measured. But even at this stage it seems clear that there are strings of objects extending in opposite directions from the central galaxy NGC 214, *and their redshifts continually increase as we approach galaxy clusters at either terminus.*

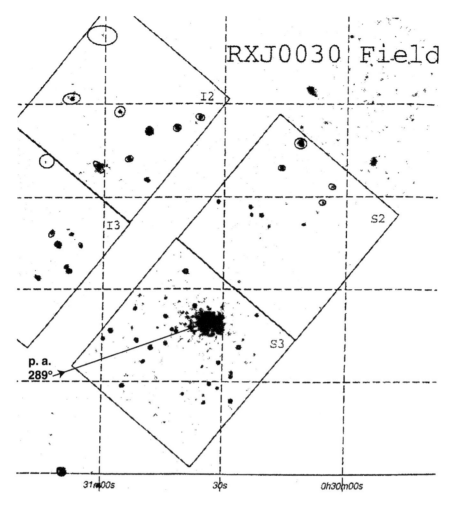

Fig 11. The Chandra observations of the RXJ0030+1628 cluster from Cappi et al. (2001) are shown. The line back to NGC 214 at p.a. = 287 deg. is indicated by the arrow.

Because of the apparent overdensity of bright X-ray sources in the region of the CRSS (or RXJ) 0030 cluster, it is extremely interesting to examine the Chandra Exposure in Fig. 11. There is also a well documented excess of faint Chandra Sources (Cappi et al. 2001). It is intriguing that the excess of Chandra X-ray sources comes from sources all about the same intensity close to the ESE side of the dominant X-ray source. In fact, these sources are all strung out in the direction back to NGC 214, about p.a. = 287 deg.. The strong X-ray galaxy itself is distended in roughly this same direction. The configuration suggests that the strong X-ray source is moving out along the line from the central NGC 214 and trailing smaller sources along its path of travel.

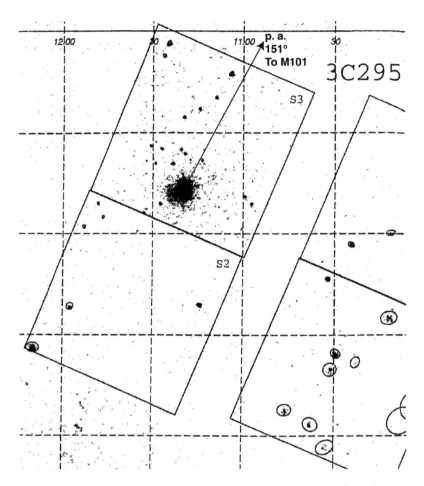

Fig. 12. The Chandra observations of the 3C295 cluster from Cappi et al. (2001) are shown. The line back to M101 at p.a. = 151 deg. is indicated by the arrow.

BACK TO M 101 WITH THE CLUSTERS 3C 295 AND ABELL 1904

By an astonishing coincidence (?) two strong X-ray galaxy clusters at around z = .5 with a reported over density of Chandra sources are presently known. The second is 3C 295 at z = .461. In Appendix A we identified it as being ejected SW from M 101 along a direction variously estimated between p.a. = 149 to 151 deg. Fig. 12 here shows that the same kind of Chandra sources trail out behind its apparent path away from M 101. Even the dense part of the X-ray image seems to be elongated roughly along this direction.

190

A further point is, of course, that the overdensities of Chandra sources around both of these X-ray clusters, found by two independent teams, have only two possible explanations. Either they turn out to have the same redshift as the central source, in which case they will have to be accepted as physical members that are all aligned with a low redshift galaxy of origin—or they turn out to have widely different redshifts, in which case the overdensity, already being interpreted as not accidental, is evidence of different redshifts physically associated long the line.

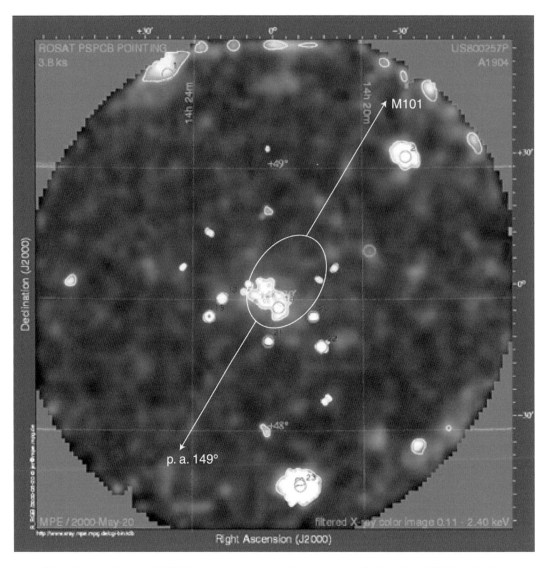

Fig. 13. A 3.8 ksec PSPC X-ray exposure of the galaxy cluster Abell 1904. Optically the cluster is elongated back toward M101 (see M101, Appendix A, Fig. 8). The relatively short X-ray exposure here shows a flattening at the head of the cluster perpendicular to the inferred motion and faint indications of X-ray material trailing back toward M101 and 3C295.

AND NOW ABELL 1904

Recall that farther out in the field we saw the Abell Cluster 1904. It is elongated along this same line from M 101, and is also a strong X-ray cluster.

We do not have high-resolution Chandra images, but we do have a relatively short PSPC exposure. Fig. 13 here shows a flattening at the head of the cluster perpendicular to the inferred line of motion and faint indications of X-ray material trailing back toward M 101 and 3C 295. The isophotes suggest an ablating central X-ray core, and it would be conclusive if observations by Chandra were to confirm this.

Given all the preceding similar cases discussed in this preliminary sampling of extended clusters, it would seem very unlikely that these alignments are due to chance. The evidence also includes multiple cases of X-ray, quasar and galaxy material originating from an active galaxy and actually travelling outward along the line of origin from the galaxy.

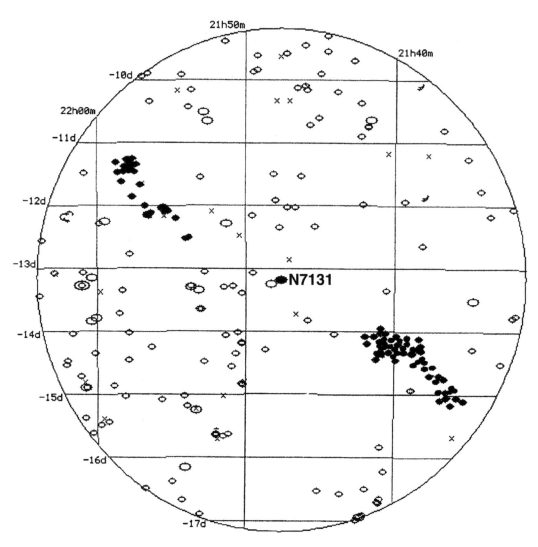

Fig. 14. This is a replot of the lines of galaxies centered on NGC 7131 that was shown in the main *Catalogue*. The SIMBAD[0] contours of individual galaxies interfere with each other because they are so densely packed. As a result, the galaxies in the line have been blacked in here for better visiblity.

THE PERFECTLY EJECTED CLUSTER

Another *Catalogue* entry is shown here in Fig. 14. Because the galaxies are packed together so closely in the Abell cluster A 2400 (z = .088) and A 2361 (z = .061) the SIMBAD plot has trouble drawing their individual outlines at this scale. In the present Figure they have been blacked in to give the shape of the dense cluster elongations. NGC 7131 (z = .018) at the center of the line is also blacked in.

The reason I suggest the characterization "perfect" is that the cluster lines are so thin and so well centered on either side of the central galaxy. They are even bent in slightly opposite directions at their ends, reminiscent of the arms in a spiral galaxy. I have long considered the spiral arms to be a result of ejection combined with rotation of the ejecting core of the galaxy (Arp 1963; 1986). I would be pleased to suggest the same explanation for the NGC 7131 configuration. In any case, the observed galaxies in Fig. 14 do not seem to call for any mathematical computations to establish the significance of the association. As to its probable cause we can make the following comments.

Summary and Interpretation

The observations considered so far suggest:

1) Jets emitting synchrotron radiation are ejected from galaxy nuclei which are in an active phase.

2) As electrons and protons travel outward they combine into atoms and evolve into compact sources that show redshifted atomic spectra.

3) When the sources impact galactic or intergalactic medium they slow and can differentiate into smaller sources.

4) Lines of quasars and AGN's generally evolve toward lower redshifts at greater separations from the parent galaxy. But lower redshift material can be entrained in the ejections and can furnish a gradient or mix of redshifts out to some higher redshift objects.

The key question that must be answered is: What causes the intrinsic redshift of the ejected material? We assume here that the only explanation which has yet to be empirically refuted is that low particle mass matter is periodically created in the active nuclei, and that the atomic transitions which give the observed redshift become more energetic as the particles age and increase in mass. (Narlikar 1977; Narlikar and Arp 1993; Arp 1998).

What would be the expected consequences of such a model? First, if a normal, hot plasma is ejected one would expect it to dissipate rather than condense. In a low mass plasma, however, particle velocities would slow in order to conserve momentum as particle masses grew. The gravitational contraction forces would also increase. We would therefore expect the plasmoid to cool and condense. (Unlike the hot gas in the Big Bang which would dissipate, and which therefore requires hypothetical cold dark matter to form galaxies). At some point in the low mass plasma, however, the ionized particles would start to combine into atoms. Initially the redshift would be high, but as time went on stars would form, luminosity would increase and the object would evolve into a normal, slightly higher redshift companion galaxy.

The trail of X-ray sources that we observed from NGC 4636 to PG 1244+026 would then represent successive ejections of low mass plasmoids which are evolving into the quasars and AGN's with redshifts that need now to be measured. The redshift of PG 1244+026 is $z = .048$; it is important to note that there are three other galaxies within 29 to 38 arcmin of NGC 4636 that have redshifts of $z = .046, .048$ and $.048$. Two of these are centered and aligned accurately across NGC 4636. It is sugggested that they represent material ejected in diffferent directions but of the same epoch as PG 1244+026.

PG 1244+026 is a strong X-ray source, an ASCA source, an infrared (IRAS) source, and has drawn 65 published references. I should make the point straightaway that although it is arbitrarily classified as an active galaxy because of an absolute magnitude which would be $M_V = -21.2$ if it were at its redshift distance, it is in fact very compact and spectroscopically continuous with the class of objects routinely called quasars. The fact that it is much more active than the other galaxies of $z = .048$ in the vicinity may reflect a larger initial quantity of ejected material, continuing material ejection in this direction, or more interaction with material in its path.

INTERACTION OF EMERGING JET WITH MEDIUM

Previous observations have shown that strong pairs of X-ray quasars tended to be ejected along the minor axes of disk galaxies. This seems natural, because in this direction an ejection would encounter the least resistance on its way out of the galaxy. It was then shown that in cases of very disrupted galaxies, the quasars tend to be smaller, more numerous and closer to the galaxy of origin. It was reasoned that if the ejection was not along the poles but through the substance of the galaxy, this would disrupt the galaxy, and slow down and break up the material being ejected. (See Arp 1998b; 1999a; 1999b.)

Aside from furnishing a satisfying explanation for the class of galaxies which were obviously very morphologically disturbed in the outer regions as a result of internal activity, this also suggested some physical characteristics of the ejected plasmoids. Qualitatively the variable mass theory would hypothecize an ionized (normal charges) plasma with near zero particle masses of relatively large cross section. This substance would be expected to interact strongly with the gas and/or gas plus magnetic field of a galaxy or intergalactic medium. In this respect it reminds one of Ambartsumian's "superfluid." In 1957, by studying the Palomar Sky Survey images, he suggested that new galaxies were born in this fashion. (See Arp 1999a, p473.) Since the mass is near zero, the initial state would be almost pure energy, and the expected initial outward velocity would be near c. This high velocity would rapidly slow in order to conserve momentum as the mass grew, and also as it pushed aginst the ambient medium.

Perhaps one of the best places to check our predictions against the observation is in the much-studied jet of M 87. (For a picture see *Seeing Red* Plate 8-18.) There some of the smaller knots have translation velocities along the jet which are very close to the speed of light (e.g., Lorentz factor 6, Biretta et al. 1999). Even though the channel along the jet must be almost completely clear of matter between the knots, there is still obvious ablation on the major knots. This implies that the plasmoids are still quite fragile at this stage and can easily be disrupted, although they are not dissipating like a hot gas. Nevertheless the initial energy of a plasmoid capable of evolving into an average-mass companion galaxy must be enough to punch through considerable galactic/intergalactic medium. In the case of M 87 only 1.4 deg. farther out, and exactly along the line of the jet, is a radio galaxy M 84 whose X-ray contours are compressed, such that it is obviously moving out directly along the line of the jet. About 4.5 deg. from the center of M 87, and again exactly along the line of the jet, is a bright quasar of redshift z = .085 with flanking pair of quasars at z = 1.28 and 1.02 (Arp 1995 and present *Catalogue* entry for PG 1211+143). As a result, we find objects of variously higher redshifts farther out along the line of the synchrotron knots in the inner M 87 jet.

PERSISTENCE OF EJECTION DIRECTIONS

In the interior of most galaxies there is probably ionized gas with embedded magnetic fields. If an ejection entrains this gas, it should elongate the magnetic lines of force along the length of the ejection. Subsequent ejections of low particle mass plasmoids would then be guided out along this same channel. If this channel stays fixed in space, as is apparently the case with the M 87 jet, it offers an explanation of how we can find

Fig. 15 (Plate 15). The Scl galaxy NGC 1232 with a +5,000 km/sec companion off the end of a spiral arm. Farther in along the same arm, near a thickening, is a small galaxy with a redshift of cz = .1. The image is from the VLT telescope of the European Southern Observatory.

lines of objects with a large spread in ages and intrinsic redshifts. Very low redshift objects (of the same order as the ejecting galaxy) may represent original material that has been entrained in the ejection.

If ejections take place in the disks of rotating disk galaxies, however, we would expect the channels to be wound up in spiral form at least to some extent. Measurements of spiral galaxies show that the magnetic fields actually do lie along the arms, and this leads to the suggestion that spiral arms are actually ejection tracks from active nuclei along which stars are being condensed. (See discussion in Arp 1963; 1969; 1986). This enables

198

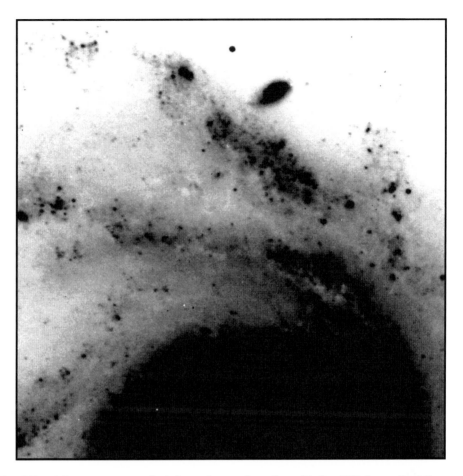

Fig. 15a. A long exposure in infrared wavelengths with the VLT shows that the $z = 0.1$ galaxy has a tail, presumably due to its interaction with the bright swelling in the spiral galaxy arm.

us to account for a baffling class of observations, namely higher redshift companions attached to the ends of spiral arms of low redshift galaxies.

New Galaxies in the Arms of NGC 1232

Fig. 15 shows that NGC 1232 has a companion of +4,900 km/sec excess redshift lying off the end of one arm, and in the same arm a disturbed region harboring a galaxy with a young stellar spectrum of redshift nearly $cz = 0.1c$ (Arp 1982; 1987). It is amusing to note that in both the NGC 4151 and NGC 1232 cases the companions were reported as having the same redshifts as the main galaxies to which they obviously belonged.

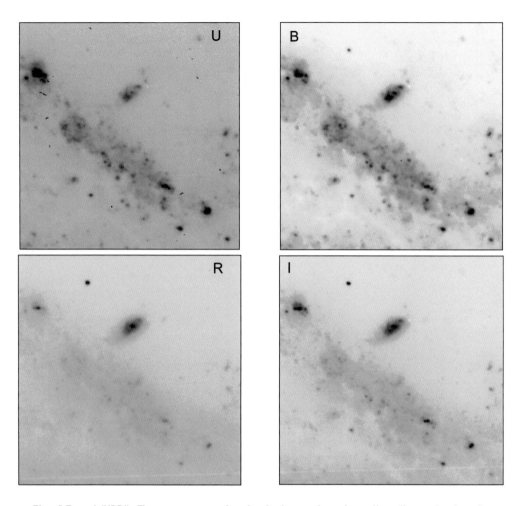

Fig. 15 a-d (UBRI). The exposures, in clockwise order, show the discordant red-shift galaxy and adjoining spiral arm in Ultraviolet, Blue, Red and Infrared wavelengths. In U and B large emission regions are seen; in longer wave-lengths the high surface brightness galaxy becomes dominated by a red nu-cleus.

In both cases it was later determined that this was a mistake, and that the companions actually had +5,700 and +4,900 km/sec higher redshifts. Naturally they were then relegated to the background! But in the case of NGC 1232B, Fig. 15 shows that an outer spiral arm leads directly to the companion. The companion and the arm have the same resolution into stars and HII regions, and this detailed picture seems to show clearly that the galaxy is an aggregate of the same material as in the arm. Following this arm back toward the center of the spiral leads to a flat section per-pendicular to the arm, which looks like the beginning of a channel of just the size along which the companion could have travelled.

Farther inward, the brightest, thickest arm section anywhere in the galaxy is encountered. The object next to this unusual section of the arm aroused my curiosity, and on one of my last runs on the Las Campanas 2.5 meter telescope I took its spectrum. It turned out to have a redshift of $cz = 28,000$ km/sec or $z = 0.1$. If it can be demonstrated that it is at the same distance as the $cz = 1,603$ km/sec main galaxy, of course, this is irrefutable proof of a discordant redshift. It had been noted earlier that it showed no sign of reddening or obscuration, which would be expected if it were seen through the dusty plane of the spiral. Because of its importance we show here a series of magnified views from the 8 meter VLT which were taken in its commissioning period, I deep (Fig. 15a *shows an interaction tail toward the thickened arm!*). In addition, lighter images in Fig. 15 UBRI (Ultraviolet, Blue, Red and Infrared) show:

a) Fig. 15 U - a faint nucleus to the galaxy and paired knots closely across it NW-SE.

b) Fig. 15 B - the knots get very large and much brighter, implying perhaps OII emission at the $z = 0.1$ redshift.

c-d) Fig. 15 R, I - the knots get fainter and the nucleus gets brighter until the I wavelength, where the object looks like a high surface brightness dwarf with a red nucleus.

The spectrum I took in 1982 showed a young galaxy. Now we see that it is morphologically disturbed. Moreover, the HII region in the anomalous redshift object would be astonishingly large at its redshift distance. It is absolutely crucial to put the slit of the spectrograph along this galaxy and register the lines and their redshifts in its various parts. There is another bright emission region in these pictures, just to the ENE of the $z = 0.1$ object. It may be similar and should also be investigated spectroscopically.

In fact there are a number of regions in the arms of this galaxy which could now be systematically investigated with large new telescopes, fast detectors and multislit spectra. I find it rather shocking that while a galaxy like NGC 1232 is often observed for its esthetic value, scientists do not even inspect the objects that are of the most fundamental importance to astronomy and physics. Detailed papers published in professional journals and in books calling attention to such objects are ignored. The indications are that we may encounter in these arms a whole new world of astronomy and physics that we have not before imagined.

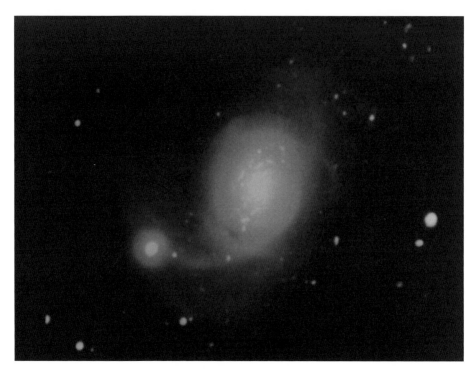

Fig. 16 (Plate 16). The active Seyfert galaxy NGC 7603 has a +8,000 km/sec companion on the end of an arm-like filament.

QUASARS IN THE ARM OF NGC 7603

Another example of high redshift objects in galaxy ejections/arms makes the previous case especially significant. NGC 7603 (*Atlas of Peculiar Galaxies* No. 92) shows a companion on the end of a spiral arm-like connection which is +8,000 km/sec relative to the parent galaxy (Fig. 16). Ironically, *after 30 years*, Fig. 17 (color) shows that the two quasar-like, high redshift objects (z = .24 and .39) have recently been measured in this same connecting filament by two young Spanish astronomers in Tenerife (López-Corredoira and Gutiérrez 2002). The results, moreover, fit well with the model we have suggested here, wherein a plasmoid entrains a trail of matter from the ejecting galaxy, is slowed and evolves into a high redshift companion. Subsequent ejections follow this magnetic tube, and we see two now in an earlier stage of evolution at z = .24 and, closest to the Seyfert, z = .39. The evidence is summarized in two color pictures included here as Fig. 16 and 17. It is remarkable that we can now support the evolutionary sequence in NGC 1232, where the redshifts drop outward along the arm from the central low redshift galaxy to z = .1 and then to .02. In NGC 7603 they go to z = .39 to .24 to .06.

202

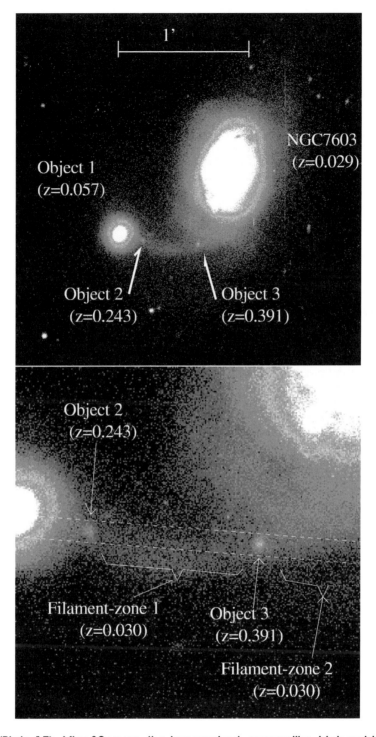

Fig. 17 (Plate 17). After 30 years, the two marked, quasar-like, high redshift objects have now been discovered in the filament from NGC 7603 with the Nordic Optical Telescope of La Palma (López-Corredoira and Gutiérrez 2002).

203

In all of the above cases we would argue that low particle mass plasmoids had been guided out along the spiral arms. There is a disturbance in the NGC 1232 arm where the $cz = 0.1$ c object lies, and this may be connected to unexplained "swellings" in the spiral arms of galaxies, as in NGC 4151, for example. (See Fig. 18a and b in the next section.) It is difficult to overestimate the rewards that might await some courageous user of modern big telescopes on these obvious targets.

The story of NGC 7603 just by itself is a poignant example of crucial scientific evidence ignored and fundamental theoretical progress suppressed. As related at the end of the Introduction to this book, Fred Hoyle in 1972 singled out the discordant companion in 1976 as forcing a crisis in Astronomy. But the latest observational evidence on the quasar like objects in the connection to this companion was rejected by several referees and journals until it was finally published in A & A. Follow-up observations on this discovery were not allowed by Allocation Committees (TAC's) on the Chandra X-ray satellite and the southern Very Large Telescope (VLT). It reminds one of large corporations whose boards of directors assure the public of large profits while the shareholders are losing large amounts of money.

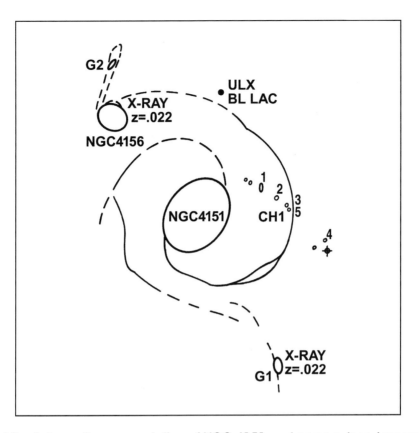

Fig. 18a. Schematic representation of NGC 4151 and companions. Low sur-face brightness features are marked with a dashed line. Galaxies in the chain west of NGC 4151 are numbered 1 through 5. The BL Lac quasar is about 4.5 arcmin N of NGC 4151 and has a redshift of $z = .615$. For more details see Arp 1977.

EJECTED OBJECTS FROM NGC 4151

One of the original active galaxies discovered by Karl Seyfert was NGC 4151. An extensive study of companion galaxies in its neighborhood was published 26 years ago (Arp 1977). Detailed morphological evi-dence was presented at that time to the effect that the companion which lay off the end of the NE spiral was in fact connected to that arm. On the end of the opposite, SW arm was a similarly connected peculiar com-panion. That was before the era of X-ray observations. *Today both are listed as X-ray galaxies with redshifts of $z = .022$ and $z = .022$.* (Fig. 18a.)

The two companions form a well aligned and centered pair about 4.7 arcmin on either side of this famous active Seyfert Galaxy. The implication

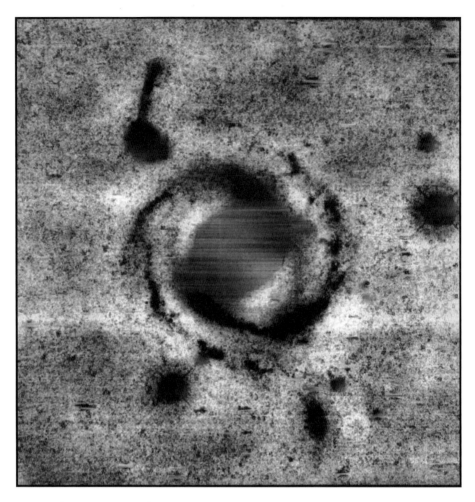

Fig.18b. Sky limited KPNO 4 meter exposures have been stacked to show faint surface brightness features in the exterior of the famous Seyfert galaxy NGC 4151. The ScI galaxy NGC 4156 to the NE and the companion to the SW have excess redshifts of +5,700 and +5,400 km/sec. Note the thickening of the arms in several places, as in NGC 1232 (Fig. 15)

is that proto galaxies were ejected in opposite directions from the active nucleus and evolved into young, higher redshift galaxies as they travelled out along, or drew out, the main spiral arms of NGC 4151. In this regard they are very much like the configuration shown in NGC 1232. There are even swellings in the arm of NGC 4151 (Fig. 18b) which resemble the thickening of the arm in NGC 1232—a thickening which is near the dis-cordant redshift galaxy of $z = 0.1$. There is a very powerful (257 X-ray counts/kilosec) BL Lac object of $z = .615$ just above a similar point of the arm in NGC 4151. Also in NGC 4151 a straight chain of galaxies leads back toward the ejecting companion NGC 4156. The chain has redshifts that range from $z = .06$ to $.24$, a configuration that can be better understood now in the context of the galaxy cluster associations presented here in the main *Catalogue of Discordant Redshift Associations*. The high redshift companions on and in spiral arms can also be better understood as associations where the ejecta have been slowed by interaction with the parent galaxy, which then slows the ejection speed and causes the galaxy to evolve close to its parent.

The alarming aspect of all of this, however, is that as the torrent of data from ever more telescopes and astronomers floods in, it is not placed in the context of previous results. In the case of NGC 4151, the X-ray nature of the two companions was treated as if these were just more background objects to be fitted to the current assumptions about a homogeneous universe. If the researchers knew about the extensive results published many years previously, they could have considered the possibility that their results, in fact, disproved those fundamental, current assumptions. For example, if the surveys of ultra luminous X-ray sources (e.g., Colbert and Ptak 2002; Roberts et al. 2001; Foschini et al. 2002) had referenced the data on BL Lac objects associated with active Seyferts (Arp 1997), it would have been obvious that the ultraluminous X-ray sources could be primarily BL Lac type quasars in the process of ejection from low redshift galaxies. (See G. Burbidge, E.M. Burbidge, H. Arp *A & A* 400, L17.)

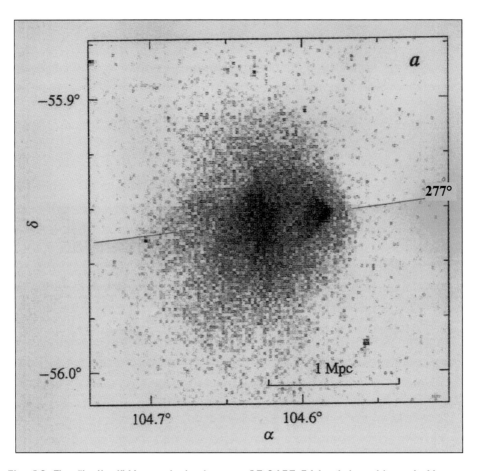

Fig. 19. The "hottest" X-ray cluster known, 1E 0657-56 is pictured here in X-rays by the Chandra satellite (Markevitch et al. 2002). The bow shock trails the "bullet," which is deduced to be colder material impinging on a warmer environment. The redshift of the galaxy cluster is 0.296!

A Bullet to a Gamma Ray Burst

In the main *Catalogue* and Fig. 1 of the M 101 Appendix we have given examples of gamma ray bursters aligned with ejection directions from active centers. We have also shown connections in high energy gamma rays between galaxies and quasars of vastly different redshifts (Intro Fig. 15 and Arp, Narlikar and Radecke 1997). This evidence strongly indicated that not only high energy X-rays, but even higher energy gamma rays appeared along the track of the ejection. The nature of the ejection, however, may have been further elucidated by the recent discovery of

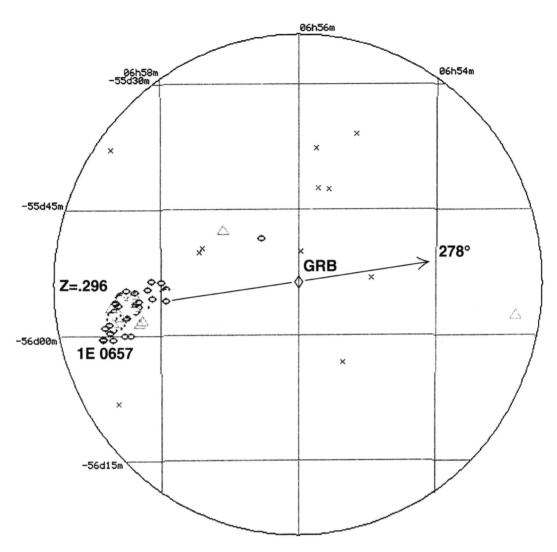

Fig. 20. At 20.8 arcmin distance from the strong X-ray cluster 1E 0657-56 is a gamma ray burster at position angle p.a. = 278 deg. The bullet emerging from the X-ray cluster is estimated to be at p.a. 277 deg., i.e., aligned within the uncertainty of the positions. The redshift of the cluster is z = .296!! Note alignment of objects to NE of cluster.

"bullets"emerging from galaxy clusters in high resolution X-ray maps by Chandra. An example is shown here in Fig. 19.

The compact X-ray source is shown leaving the center of the cluster trailing a wake of interaction behind it. Fig. 20 illustrates that only 20.8 arcmin away is a recorded gamma ray burst (10082a in *ApJ* 563, 80) at position angle 278 deg. The position angle of departure of the bullet from the

209

galaxy cluster is 277 deg. Within the uncertainty in the gamma ray position and the uncertainty of the center of the cluster, the directions are essentially coincident. It appears that the X-ray ejecta has recently produced a gamma ray outburst, as more famous quasars like 3C279 are known to do.

What is intensely interesting about the wake that the bullet is leaving is that, as the discoverers of this phenomenon have emphasized (see section 5 here on Abell 3667), the outgoing material is colder than the medium it is passing through. Thus a very compact, cold plasmoid is emerging from some central object. If it is to evolve into a quasar, and then a galaxy, this seed must exist in a low entropy, almost massless state. The charged, initially low particle masses could be frozen into this state by tangled magnetic fields in their environment, and guided by directional fields. The initial energy of ejection, however, has to be large enough to permit escape into the field.

It is intriguing to compare this empirical conclusion with conventional postulates of cold dark matter (necessary to condense galaxies) and dark energy (necessary to explain redshift discrepancies with conventional distances). Cosmologists have had to assume that these undetected substances were created in the Big Bang and that they permeate the current universe. In the observations presented here, we would argue that we are actually witnessing the birth (or perhaps reemergence) of matter, which is evidently in a cold state. We can then infer that the charged particles combine into atoms, and observe the sequence of their physical forms as a function of time. In my opinion the Machian type theory of Hoyle and Narlikar with variable particle masses (see end of Introduction) would then be more physically relevant than the current General Relativity theory.

210

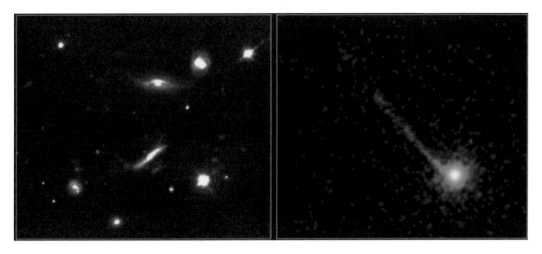

Fig. 21. On the right is the Chandra, X-ray image of the quasar J113007.-144927 (V = 17.2, z = 1.187) showing the "longest X-ray jet detected at high redshift" (Siemiginowska et al. 2002). On the left is a Space Telescope (HST) image, on the same scale, of an interacting quartet of galaxies at the tip of the "jet."

A Quasar with an Ablation Tail

Fig. 21 presents, on the right, a Chandra X-ray map of a bright, very active quasar. It shows what the discoverers call the "longest X-ray jet detected at high redshift" (Siemiginowska et al. 2002). But the HST image on the left shows at least four very disturbed galaxies just at the tip of the jet. I would suggest here that it is not a jet at all, but a trail of material ablated from the quasar as it passed through the denser medium associated with the interacting group.

In support of this interpretation the following arguments can be made:

1) The interacting group of galaxies is very unusual, and there would be an extremely small chance of them falling accidentally at just the beginning of the X-ray trail.

2) The "jet" is completely one sided.

3) It originates at the far boundary of the group (Fig. 22).

4) It is broader at its origin at the quasar and narrows down the tail.

5) The X-ray contours are compressed at the head of the quasar in the direction of the motion (see Fig. 23).

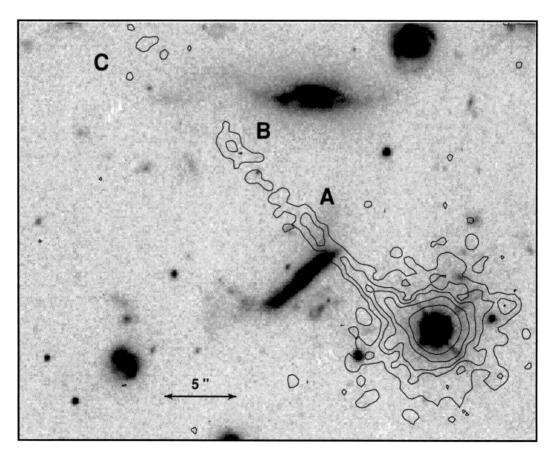

Fig. 22. Overlay of the X-ray contours on the optical image showing relation of X-ray tail to the interacting quartet at $z = .312$ (Siemiginowska et al. 2002).

6) *The trail of material does not originate at the center of the quasar image, but enters it slightly, but definitely to the SSE of center.*

7) *The relatively lower fequency radio emission at 1.4 GHz trails behind each of the X-ray knots A, B, C indicating the older radio plasma is being stripped as the X-ray knots move through the medium in the direction of the quasar. (Fig. 23)*

I would attribute the small displacement of the tail from the center of the quasar to a small counterclockwise rotation of the quasar spinning the material off to the side as it ablates (like throwing a curve in baseball).

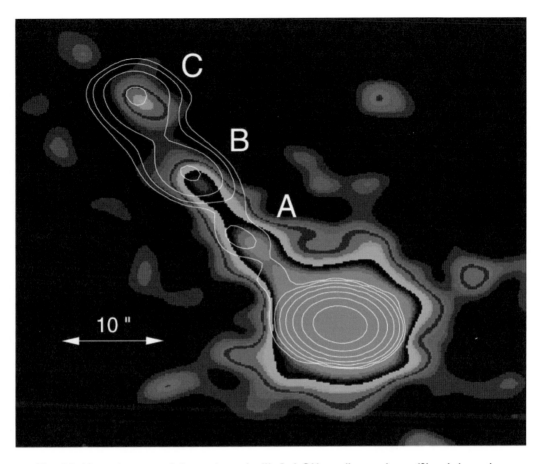

Fig. 23. X-ray (grey scale) overlayed with 1.4 GHz radio contours (Siemiginowska et al. 2002) showing displacement of radio knots downstream from the X-ray knots in the ablating quasar.

If we adopt the low particle mass plasmoid model for the quasar, then for the first time this object might enable us to make some estimates of density and cohesion properties of the substance from which the galaxy/quasar is actually evolving, and perhaps, very importantly, the slowing effect of the intergalactic medium on ejected quasars. I find that prospect very exciting.

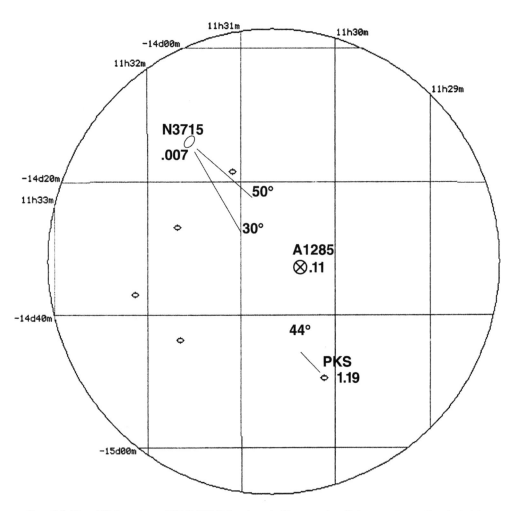

Fig. 24. The IRAS galaxy NGC3715 is about 40 arcmin distance from the bright PKS quasar. The position angle of the minor axis of the galaxy is p.a. = 50 deg.; the direction from the quasar to the galaxy is p.a. = 30 deg.; and its tail is at p.a. = 44 deg. Abell 1285 is a strong X-ray cluster.

Fig. 24 shows that the quasar is only 40 arcmin from a large galaxy (NGC 3715, z = .007). It is a typical configuration, as we have seen here in many cases in the main *Catalogue of Discordant Redshift Associations*, and I would interpret it as an ejected quasar. At this distance from the main galaxy it has run through a group of earlier ejected galaxies. (The redshift of the group is .312, or in the reference frame of the galaxy, z = .302, essentially exactly on a principal quantized redshift peak.) Hydrogen absorption lines of this redshift are identified in the quasar, showing that the HI is spread over the area of the group and beyond. A *strong X-ray cluster of galaxies* (Abell 1285 at z = .105) extends generally SW in the direction of the quasar from the NGC galaxy.

Near the end of the preceding Appendix A, the conventional large distances found for galaxy clusters from the S-Z effect are discussed in relation to the much smaller distances obtained from the observations presented throughout the current *Catalogue*. We will now attempt to sketch a plausible scenario for the formation of nearby clusters of lower luminosity galaxies.

A MECHANISM TO PRODUCE CLUSTERS OF GALAXIES

The best example of quasars that have been stopped near a very disrupted galaxy is the group of 18 quasar/AGN's of $2.389 \leq z \leq 2.397$ only 1.5 arcmin S of the extremely disrupted blue radio galaxy 53W003 (Keel et al. 1998; Keel et al. 1999; Arp 1999b). In this unusual group the objects are faint, suggesting that a larger object may have broken into smaller objects. Most noteworthy is his regard is that some of the broad line quasars are not point sources, but HST images show them as almost comet shaped objects—surely a sign of unusually strong interaction with a medium. The Keel et al. (1999) analysis makes a very cogent remark: "The objects from our emission candidate list that have been spectroscopically confirmed are all obvious AGNs, with a mix of broad- and narrow-lined cases." Since the broad line-narrow line criterion is usually used to separate quasar from galaxy spectra, this suggests that if the material in a normal broad line quasar is spread out, one will get naturally lower gas densities and hence narrow lines. The difference between a quasar and a galaxy of the same redshift would then be due only to its interaction history.

To carry this thought forward to the elongated clusters of galaxies we have discussed in the present paper, we have remarked that a low particle mass plasmoid could differentiate into smaller pieces when interacting with a medium or normal mass clouds. In fact an encounter with a medium of small cloudlets might separate a plasmoid into many smaller units. These smaller, less compact units would naturally be strung out in a line in the direction of ejection. We might conclude that this is a possible way of turning an ejected X-ray quasar, or proto quasar, into an elongated cluster of X-ray galaxies. *Since the three principle kinds of extragalactic X-ray objects known are X-ray galaxies, X-ray quasars and X-ray*

215

clusters, it would be very satisfying to unify the class by positing that the X-ray ejections come out of the dense, energetic nuclei of galaxies and evolve into the emission line, X-ray quasars and the non-equilibrium, active clusters of galaxies.

One of the most vexing problems in the subject of galaxy clusters has been "cooling flows." Why is the strongly radiating gas not cooler near the center of the clusters? After many complicated models it has finally been realized that most of these clusters have an active AGN at their centers. The irony here is that the plasmoids intermittently ejected from these central AGNs must be cool if we are to identify low particle masses as the cause of the high redshifts of the proto quasar/galaxies. (The charged particles would initially be frozen into the plasmoid by the magnetic fields.) But the evidence developed in this *Catalogue* would suggest that the kinetic energy of the ejected plasmoids could heat the intracluster medium. The tangential arcs observed in clusters then would then be associated with compression shells generated in explosions from a spherical nucleus during the process of intermittent ejections.

REFERENCES

Arp, H. 1963, "On the Evolution of Galaxies," *Scientific American*, 208, 70

Arp, H. 1968, *Astrofizika* 4, 59

Arp, H. 1969, "On the Origin of Siral Arms in Galaxies," *Sky and Telescope* p. 385, December 1969

Arp, H. 1977, *ApJ* 218, 70

Arp, H. 1982, *ApJ* 263,54

Arp, H. 1986, *IEEE Transactions on Plasma Science*, 14, 748

Arp, H. 1987, *Quasars, Redshifts and Controversies*, Interstellar Media, Berkeley

Arp, H. 1995, *A & A* 296, L5

Arp, H. 1996, *A & A* 316,57

Arp, H. 1997, *A & A* 319, 33

Arp, H. 1998a, *Seeing Red: Redshifts, Cosmology and Academic Science*, Apeiron, Montreal

Arp, H. 1998b, *ApJ*496,661

Arp, H. 1999a, *IAU Symp* 183, 290 and 473

Arp, H. 1999b, *ApJ* 525,594

Arp, H. 2001, *ApJ* 549, 780

Arp and Russell. 2001, *ApJ* 549, 802

Arp, H, Narlikar, J., Radecke, H.-D. 1997, *Astroparticle Physics* 6, 387

Biretta, J. A.; Perlman, E.; Sparks, W. B.; Macchetto, F. 1999, *The Radio Galaxy M 87*, eds. Röser, H-J., Meisenheimer, Springer, 210

Böhringer, H. et al. 2000, *ApJS* 129,435

Buote, D., Canizares, C. 1966, *ApJ* 457, 565

Brandt, W., Hornschemeier, A., Schneider, D., et al. 2000, *AJ* 119, 2349

Cappi, M., Mazzotta. P., Elvis, M., et al. 2001, *ApJ* 548, 624

Colbert, E. Ptak, F. 2002, *ApJS* 143, 25

Crawford, C., Allen, S., Ebeling, H., Edge, A., Fabian, 1999, *MNRAS* 306, 857

Ebeling, H., Edge, A. C., Allen, S. W., Crawford, C. S., Fabian, A. C., Huchra, J. P. 2000, *MNRAS* 318, 333

Foschini, L., Di Cocco, G., Ho, L., et al. 2002, astro-ph/0206418, /0209298

Jones, C.; Forman, W.; Vikhlinin, A.; Markevitch, M.; David, L.; Warmflash, A.; Murray, S.; Nulsen, P. 2002, *ApJ* 567, L115

Keel, W., Windhorst, R. Cohen, S., Pascarelle, S., Holmes, M. 1998, *NOAO Newsletter* 53, 1

Keel, W., Cohen, S., Windhorst, R., Waddington, I. 1999, *AJ* 118, 254

López-Corredoira, M. Gutiérrez, C. 2002, astro-ph/0203466 and *A&A* 390, 15L

Markevitch, M., Gonzalez, A., David, L., Vikhlinin A., Murray S., Forman, W., Jones C., Tucker, W. 2002, *ApJ* 567, 27

Matsushita, Kyoko; Makishima, Kazuo; Ikebe, Yasushi; Rokutanda, Etsuko; Yamasaki, Noriko; Ohashi, Takaya 1998, *ApJ* 499, L13

Narlikar, J. 1977, *Ann. Physics* 107, 325.

Narlikar, J., Arp, H. 1993, *ApJ* 405, 51

Pietsch, W., Arp, H. 2001, *A & A* 376, 393

Radecke, H.-D. 1997, *A & A* 319, 18

Roberts, T., Goad, M. and Ward, M. 2001, *MNRAS.* 325, L7

Rosati, P., Lidman, C., della Ceca, R.; Borgani, S., Lombardi, M., Stanford, S. A., Eisenhardt, P. R., Squires, G., Giacconi, R., Norman, C. 2000, *Messenger* 99, 26

Röser, H-J., Meisenheimer, K. 1997, *The Radio Galaxy M 87*, Springer

Siemiginowska, A., Bechtold, J., Aldcroft, T., Elvis, M., Harris, D., Dobrzycki, A. 2002, *ApJ* 570, 543

Stockton, A., Ridgeway, S. 1998, *AJ* 115

Struble, M., Rood H. 1999, *ApJS*, 125, 35

Vikhlinin, A., Markevitch, M., Murray, S. 2001, *ApJ* 551,160

Wilson et al. 2001, *ApJ* 546

Worral, D., Birkinshaw, M., Hardcastle, M. 2001, *MNRAS.*

Glossary

Absolute magnitude	The brightness (measured in magnitudes) that an object would have if observed from a distance of 10 parsecs (32.6 light years).
Absorption line	Energy missing from the spectrum of an object in a narrow range of wavelengths, owing to absorption by the atoms of a particular element. The spectrum shows a black line where a characteristic color line would appear in case of emission of the same wavelength by the atoms.
Active galaxy	A galaxy with extremely high emission of radiation especially in the high-energy range: UV-radiation, X-rays, Gamma rays. Well-known examples are Seyfert galaxies, Markarian galaxies, radio galaxies, BL Lac objects, and quasars.
A posteriori probability	The probability, after an event has occurred, that it would occur.
Apparent magnitude	The brightness that an object appears to have at its actual distance, measured in magnitudes. (The faintest stars visible to the unaided eye are about 6th magnitude, and the faintest stars and galaxies photographed in large telescopes are about 30th magnitude).
A priori probability	The probability, before an event has occurred, that it will occur.
Barred spiral	A spiral galaxy in which the spiral arms unwind from a spindle shaped "bar" of stars that forms the galaxy's inner region.
Beppo Sax	X-ray astronomy satellite initiated by the Italian Space Agency
Big Bang theory	The theory that the universe began its expansion at a particular point in space-time.
BL Lac objects	Objects with spectra dominated by non-thermal, continuum radiation. Morphologically a transition between quasars and galaxies. Marked by very strong radio and X-ray emission.
BSO	Blue stellar (appearing) object.
Bremsstrahlung	Radiation emitted by a charged particle which is de- accelerated if it encounters an atom, molecule, ion *etc.*
CCD	"Charge coupled device": Light-sensitive electronic chips used in modern astronomy to record and to measure the light received.
Chain of galaxies	A group of four or more galaxies that roughly form a line on the sky.
Chandra	X-ray astronomy satellite launched by NSA with highest resolution
Compact source	A region emitting large amounts of visible, radio or X-ray energy from a small apparent area on the sky.
Companion galaxies	Smaller galaxies accompanying a large, dominant galaxy in a galaxy pair or group.
CRSS	Cambridge ROSAT Serendipity Survey

cz	Redshift expressed in units of the speed of light (c = 300,000 km/sec)
Dark matter	Matter invisible to present astronomical instruments.
Deconvolution	A mathematical operation which helps to restore the true characteristics of an observed object. If the influence of the instrument (*e.g.* the point spread function) is known, the process allows the actual shape and intensity of the object to be better seen.
Declination	An angular positional coordinate of astronomical objects, varying from 0 degrees at the celestial equator to 90 degrees at the celestial poles.
Discordant redshifts	Redshifts which are other than expected at the distance of the object.
Dss2	Digital sky survey (from ESO)
Δz	The difference between two redshifts: $z_1 - z_2 = \Delta z$.
Electron	An elementary charged particle, a constituent of all atoms, with one unit of negative electric charge.
Emission line	A "spike" of excess energy within a narrow wavelength range of a spectrum, typically the result of emission of photons from a particular type of atom in an excited state.
ESO	European Southern Observatory
Excited state	An orbital state of an atom in which at least one electron occupies an orbit larger than the smallest allowed orbits. If an electron jumps to a lower orbit it emits a photon with an energy characteristic of the separation of both orbits. The result is an emission line in the spectrum of the atom.
EXSS	Extended X-ray Source Survey. From Einstein satellite observations, usually indicates clusters of X-ray galaxies.
Frequency (of radiation)	The number of times per second that the photons in a stream of photons oscillate, measured in units of hertz or cycles per second.
Galactic rotation	The collective orbital motion of material in the plane of a spiral galaxy around the galactic center.
Galaxy	An aggregate of stars and other material which forms an apparently isolated unit in space, much larger than star clusters (which are normal constituents of galaxies).
Gamma rays	A particular type of electromagnetic radiation of very high frequency and very short wavelength. Its origin is processes within the nucleus of an atom.
Globular cluster	A star cluster of spherical shape containing up to several 100 000 stars of very high age. Globular clusters form a spherical halo around the Milky Way and other galaxies.
Gravitational lens	An object with a large mass that bends the paths of photons passing close to it.
H_0	The Hubble constant defined as the ratio of a galaxy's redshift to its distance (distance often estimated from its apparent magnitude); its value is generally quoted as $H_0 = 50$ to 100 km s^{-1} Mpc^{-1}.

HI	Neutral (non-ionized) hydrogen, usually observed by radio telescopes, which detect the radio emission arising from the transition between different states of spin alignment of the atom's electron and the proton in its nucleus.
HII region	A gaseous clump of predominantly ionized hydrogen, excited by young, hot stars within it, and which therefore shows conspicuous emission lines.
HRI	High Resolution Instrument on the ROSAT X-ray Telescope
Hubble constant	See H_0
Isophotes	Lines connecting points of equal intensity on a skymap.
Jet	A linear feature, much longer than it is wide, usually straight, and inferred to arise from collimated ejection of material.
Karlsson formula	The mathematical expression for the peaks observed in the redshift distribution of quasars. $(1 + z_{i+1}) = 1.23(1 + z_i)$ or $\Delta\log(1 + z) = .089$.
Late type galaxies	Galaxies showing a rotational disk and increasing amounts of young star population.
Light year	The distance light travels in one year, approximately 6 trillion miles or 10 trillion kilometers.
Local Group	The small cluster of about 20 galaxies that includes our Milky Way and the giant spiral (Sb) galaxy, the Andromeda Nebula (M31).
Local Supercluster	The largest nearby aggregation of groups and smaller clusters of galaxies, with the rich Virgo Cluster of galaxies near its center.
Luminosity class	Classification scheme of stars according to their luminosity. It extends from class I for supergiants to class VI for white dwarfs. Also for galaxies I-IV with I having the best-defined spiral arms.
M	"Messier." A catalogue of nebulae, clusters, and galaxies compiled by Ch. Messier in 1784 (e.g. M87).
Magnitude	A measure of objects' brightness in which an increase by one magnitude indicates a decrease in brightness by a factor of 2.512.
Markarian galaxy	Galaxy with stron ultraviolet emission discovered in surveys by the Armenian astronomer B. E. Markarian = Infrared Astronomical Satellite measuring apparent magnitudes of sources in the infrared at 12, 25, 60 and 100 microns.
MERLIN	Multi-Element Radio Linked Interferometer Network at Jodrell Bank, England
Milky Way	Our own galaxy, a spiral galaxy in the Local Group of galaxies.
Minor axis	The axis about which a galaxy rotates. (It is perpendicular to the disk).
Mpc (megaparsec)	One million parsecs.
NGC	"New General Catalogue of Nebulae and Clusters of Stars." A catalogue published in 1888 by J. Dreyer. It contained 7840 star clusters, nebulae, and galaxies. Appendices (called IC = Index Catalogue) extended it to more than 13,000 objects.

Objective prism	A wedge-shaped glass that provides small spectra of an entire field of bright sources.
Parsec	A unit of distance, equal to 3.26 light years.
Peculiar galaxy	A galaxy which does not have the standard, symmetrical form of most galaxies.
Photon	The elementary particle that constitutes light waves and all other types of electromagnetic radiation.
Plasmoid	A dense cloud of ionized particles.
Point Spread Function (PSF)	The mathematical function describing now the light of a point-like source is spread out while passing through an astronomical instrument. The resulting image is not a point but a small disk whose radius is determined by the interaction of the radiation with the instrument.
Probability of association	If no physical association exists between objects, the probability that an observed configuration is a chance occurrence. Technically it is one minus the chance probability.
PSPC	ROSAT photon counting X-ray telescope with a field about 55 arcmin.
Quantization	The property of existing only at certain, discrete values.
Quasar	A pointlike source of light with a large redshift, often a source of radio and X-ray emission as well.
Radio lobe	Radio emission from appreciably extended areas on either side of a galaxy, often connected to the galactic nucleus by a radio-emitting jet.
Radio source	An astronomical object that emits significant amounts of radio waves.
Redshift	The fractional amount by which features in the spectra of astronomical objects are shifted to longer (redder) wavelengths.
Redshift-distance law	The hypothesis that an object's distance from us is proportional to its redshift (the usual interpretation of Hubble's law).
Redshift periodicity	The tendency of observed redshifts to occur with certain values at certain well-defined intervals from one another.
REX	Radio X-ray source
ROSAT	The German built, X-ray (*Röntgen*) Telescope
Right ascension	An angular coordinate of an astronomical object, measured eastward around the celestial equator (0 to 24 Hours) from the vernal equinox.
Schmidt telescope	A telescope with both a reflecting mirror and a correcting plate which can photograph a relatively large portion on the sky without distortion.
Seyfert galaxy	A special type of active galaxy (mostly spirals) detected by C. Seyfert. Seyfert galaxies are characterized by extremely bright cores whose luminosity shows extensive variability. They are also bright in infrared radiation and X-rays.
Spectrum	The intensity of light from an object at each wavelength observed, using a prism or a grating. The result is a sequence of colored lines or strips characteristic of the chemical elements that emit the light.

222

Spiral galaxy	A galaxy in which the bright stars and interstellar gas and dust are arranged in a rotating, flattened disk within which prominent spiral arms of young stars and H II regions are visible. Spiral galaxies are classified as Sa, Sb, Sc (or SBa, SBb, SBc... if they are barred spirals). This sequence represents decreasing diameters of the central bulge and increasing separation of the individual spiral arms. An "I" added to the classification supposedly indicates high luminosity.
Starburst galaxy	A galaxy with an exceptionally high rate of star formation.
Supernova	An exploding star, which becomes (temporarily) thousands of times more luminous than the brightest normal star in galaxy.
Synchroton radiation	Radiation emitted by charged particles moving at nearly the speed of light whose trajectories are bent in a magnetic field.
Tully-Fisher distance	For rotating galaxies, a correlation found by R.B. Tully and J.R. Fisher between the luminosity and the width of the 21-cm radio line. It allows, in principle, to estimate the mass and hence the luminosity of a galaxy from the profile of its 21-cm line, and hence its distance.
Universe	All observable or potentially observable matter that exists.
Wavelength	The distance between two successive wave crests in a series of sinusoidal oscillations.
X-rays	A particular type of electromagnetic radiation, of high frequency and short wavelength.
X-ray source	An astronomical object that emits significant amounts of X-rays.
z	The symbol for redshift, defined as the displacement of spectral features in wavelength, expressed as a fraction of the original wavelength $z = \Delta\lambda/\lambda$.
z_o	Measured redshift transformed into the rest frame of the parent galaxy, i.e., $(1 + z_0) = (1 + z)/(1 + z_{gal})$.

Index of Objects

Colour Plates

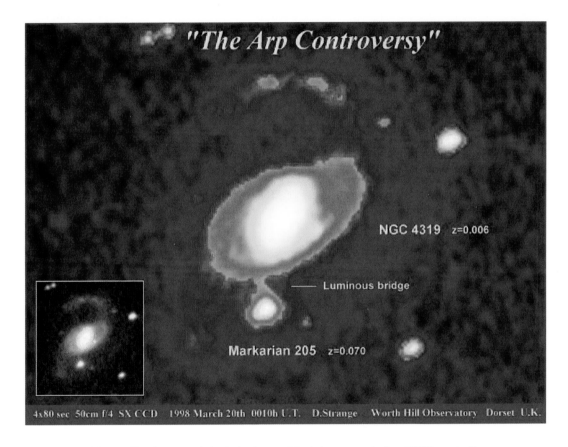

Plate 13 Intro - The famous debate between big telescopes in the 1970's as to the reality of the connection between the galaxy NGC 4319 and the quasar/AGN Markarian 205 has been settled by these CCD frames taken by D. Strange with a 50 cm telescope in the English countryside.

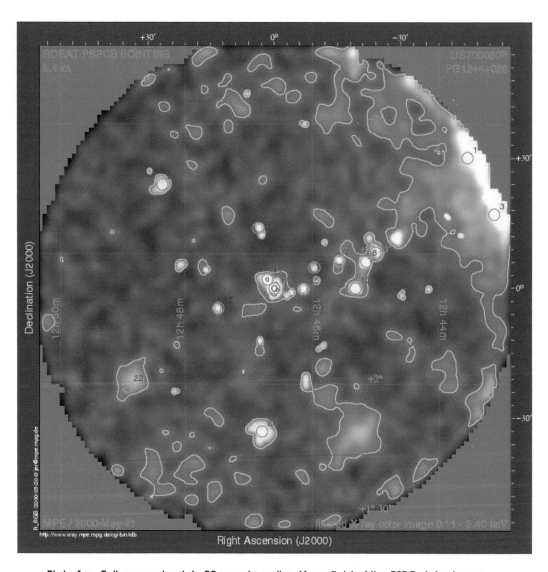

Plate 1a - Full, approximately 50 arcmin radius, X-ray field of the PSPC detector on ROSAT. A line of bright compact X-ray sources connects NGC 4636 (z = .003) at upper right with the object in the center, the AGN PG1244+026 (z = .048). (See Fig. 1b.)

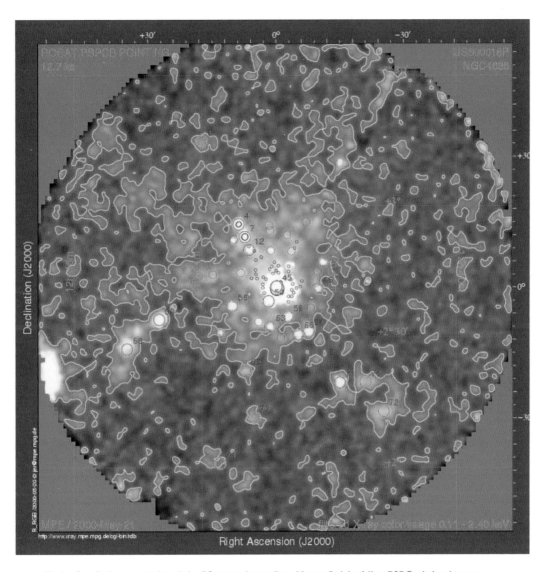

Plate 2a. Full, approximately 50 arcmin radius, X-ray field of the PSPC detector on ROSAT. A line of bright compact X-ray sources extends in either direction from NGC 4636 (a bright X-ray, E galaxy with z = .003). (See Fig. 2b.)

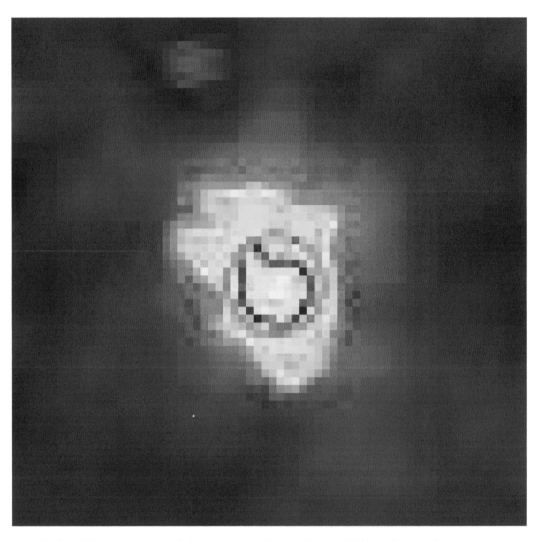

Plate 4. This is an enlarged view from the high resolution, HRI X-ray image. X-ray material appears to emerge from this central galaxy and extend out along the p.a. = 42 deg. line back toward NGC 7541.

Plate 7. An optical image, with X-ray contours superposed, of the cluster RXJ0152.7-1357 at z = .83. At its redshift distance it would be one of the most luminous clusters known. It points, however, nearly at the strong X-ray E galaxy NGC 720. (See next Figure.)

Plate 15. The ScI galaxy NGC 1232 with a +5,000 km/sec companion off the end of a spiral arm. Further in along the same arm, near a thickening, is a small galaxy with a redshift of cz = .1. The image is from the VLT telescope of the European Southern Observatory.

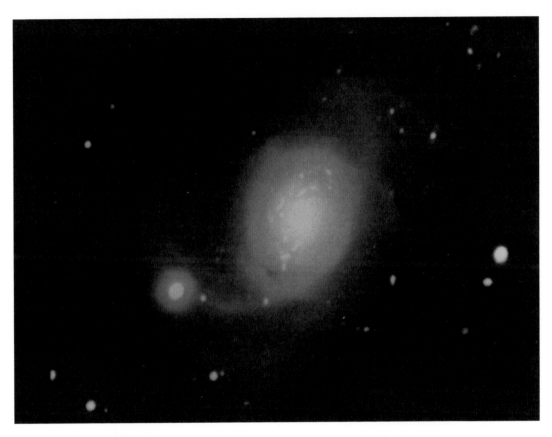

Plate 16. The active Seyfert galaxy NGC 7603 has a +8,000 km/sec companion on the end of an arm-like filament.

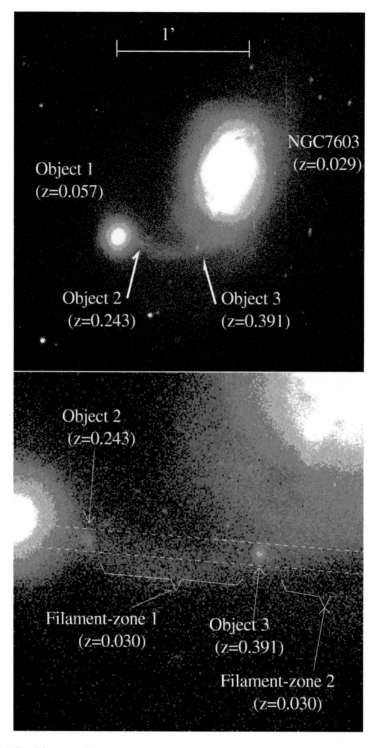

Plate 17. After 30 years, the two marked, quasar-like, high redshift objects have now been discovered in the filament from NGC 7603 with the Nordic Optical Telescope of La Palma (López-Corredoira and Gutiérrez 2002).

234

20765593R00132

Printed in Great Britain
by Amazon